CEN Q & A Handbook Vol III

CEN Q & A Handbook Vol III

Mark Boswell

MSN, FNP-C, CEN, CFRN, CTRN, CPEN, TCRN, SCRN, NR-P, W-EMT, EMT-T

Boswell Emergency Medical Education Technology
2017

Copyright © 2017 by Boswell Emergency Medical Education Technology

All rights reserved. This book or any portion thereof may not be reproduced or used in any manner whatsoever without the express written permission of the publisher except for the use of brief quotations in a book review or scholarly journal.

First Printing: 2017

ISBN 978-1-365-82816-4

Boswell Emergency Medical Education Technology
Simpsonville, SC

www.PassTheCEN.com

Introduction

This is the third volume of the CEN Review Handbook series.

This volume continues the same method and outline as set forth in the prior two volumes.

Included in this publishing is a stand-alone 150 question practice CEN-Like exam. Complete answers with rationales are provided in Part 2 of this handbook.

The distribution, skill level, theme, and tone of the questions presented here attempt to mirror the actual CEN exam as much as possible.

The reader or CEN candidate is reminded that it is in repetition of question and answering that the cognitive knowledge base is built and reinforced.

It is my hope that this volume becomes one of the tools to assist in your CEN exam preparation.

As always, feel free to contact me with any feedback, comments, questions or concerns. I strive to be easily reachable by my students and networks.

Good luck!

-Mark Boswell
MSN, FNP-C, CEN, CFRN, CTRN, CFRN, CPEN, TCRN, SCRN, NR-P, W-EMT, EMT-T

Boswell Emergency Medical Education Technology
www.PassTheCEN.com
facebook.com/bemetweb
BoswellEMT@Gmail.com

Part 1: Questions

1) Peritoneal lavage may be considered as a possible diagnostic tool in the evaluation for abdominal injuries. Which of the following is a contraindication to performing this?

   a) an allergy to contrast dye
   b) a distended bladder
   c) an unresponsive patient
   d) recent abdominal surgery within the last 90 days

2) Acute abdominal pain in the geriatric patient is most likely to be associated with which of the following conditions?

   a) bowel obstruction and intussusception
   b) diverticulitis
   c) inflammatory bowel disease
   d) ulcerative colitis

3) Which of the following is least likely associated with a stomach injury?

   a) bowel sounds in the chest
   b) blood in the naso-gastric suction contents
   c) absent bowel sounds
   d) epigastric tenderness

4) A patient is being treated for upper GI bleeding. They have a history of liver disease. During this patient's care which is the most likely expected drug that would be given via a nasogastric tube as part of their treatment?

    a) Propylthiouracil (PTU)
    b) Vasopressin (Pitressin)
    c) Heparin
    d) Magnesium citrate

5) Which of the following findings is most likely associated with a presentation of choleycystitis?

    a) Murphy's sign
    b) Lhermitte's sign
    c) Cullen's sign
    d) Psoas sign

6) What are the most common causes of acute pancreatitis?

    a) trauma and alcohol abuse
    b) hypercalcemia and drug abuse
    c) alcohol abuse and biliary tract disease
    d) trauma and post-operative syndrome

7) Which of the following conditions is most likely to have a fluid volume deficit?

    a) pancreatitis
    b) appendicitis
    c) gastric ulcer
    d) choleycystitis

8) Which of the following symptoms is most likely associated with a mesenteric infarction?

   a) constipation
   b) abdominal distention and free air
   c) vomiting fecal-like material
   d) abdominal distention and bloody diarrhea

9) Which of the following findings is most specific for pericarditis?

   a) presence of a third heart sound (S3)
   b) bilateral crackles
   c) pericardial friction rub
   d) chest pain unrelieved by position change

10) Following administration of Lidocaine for nonsustained ventricular tachycardia, which of the following symptoms is most likely associated with Lidocaine toxicity?

   a) seizures
   b) cushings response
   c) tachycardia
   d) hypertension

11) Which injury pattern is most commonly associated with aortic trauma?

   a) penetrating chest trauma
   b) deceleration forces and shearing
   c) blunt chest trauma
   d) chest trauma with associated rib fractures

12) Which of the following findings is most consistent with a ruptured descending aorta?

a) distended jugular veins and muffled heart sounds
b) pulse amplitude greater in the arms than the legs
c) pulse amplitude greater in the legs than the arms
d) blood pressure discrepancy in right versus left arms

13) A patient presents with substernal chest pain and shortness of breath. You have given nitroglycerin sublingual x3 and oxygen. These have not helped the pain. However 5mg of Morphine IV x1 has reduced the pain some. Which of the following is the most likely underlying condition causing this response?

a) angina pectoris
b) pericarditis
c) heart failure
d) acute myocardial infarction

14) Which of the following is potentially a reversible cause of ventricular fibrillation (VF)?

a) digitalis toxicity
b) hyperthermia
c) hypervolemia
d) hypothermia

15) What is the purpose of performing cardiac defibrillation?

a) to cause a temporary asystole
b) to synchronize the rhythm
c) to produce a sinus rhythm
d) to "jump start" the heart

16) Which of the following drugs is contraindicated in a wide-complex tachycardia?

a) verapamil
b) lidocaine
c) diltiazem
d) procainamide

17) What is the purpose of giving beta-adrenergic blockers to the patient experiencing a myocardial infarction?

a) to reduce blood pressure
b) to increase catecholamine levels
c) to reduce myocardial oxygen consumption
d) to block parasympathetic activity

18) Your patient's potassium result is 7.5 mEq/L. Which of the following are most likely to be seen?

a) dradycardia and a narrow QRS
b) tachycardia, peaked T waves and a wide QRS
c) bradycardia, peaked T waves and a wide QRS
d) tachycardia and a narrow QRS

19) Occlusion of the left coronary artery will most likely result in damage to which area?

a) right ventricle
b) posterior wall of the heart
c) anterior wall of the heart
d) inferior wall of the heart

20) Which of the following fibrinolytics is most likely to cause an allergic reaction?

a) Reteplase (Retavase)
b) Streptokinase (Streptase)
c) Tenecteplase (TNKase)
d) Alteplase (Activase)

21) Which of the following terms is the correct name for this "the force with which a cardiac chamber must eject blood during systole"?

a) afterload
b) stroke Volume
c) systemic vascular resistance
d) preload

22) Which of the following is most specific for left sided heart failure in the adult patient?

a) systolic murmur
b) S3 or ventricular gallop
c) diastolic murmur
d) S1 heart sound

23) Which of following regions of the brain is responsible for respiratory and cardiac systems?

a) hypothalmus
b) diencephalon
c) frontal Lobe
d) medulla

24) A patient is complaining of head and neck pain and does not recall the events prior to the assessment. His GCS is 14 and a hematoma is visualized over the occiput. Which of the following is the most appropriate intervention?

a) perform a complete head to toe assessment.
b) obtain alcohol and toxicology panels
c) apply c-spine immobilization
d) administer prescribed pain medicine

25) Which of the following are most specifically related to a spinal cord injury?

a) dxophthalmos
b) contaminated CSF on lumbar puncture
c) hypertension with tachycardia
d) paresthesias of the extremities

26) Which of the following is most specific to increasing intracranial pressure (ICP)?

a) hypertension
b) bradycardia
c) decrebrate posturing
d) unequal pupils

27) What is the primary purpose of giving diuretics in acute heart failure?

a) to decrease preload
b) to decrease afterload
c) to increase cardiac contractility
d) to decrease myocardial oxygen demand

28) What level of protection is appropriate to wear when working with a patient exposed to a highly contagious disease associated with an act of bioterrorism?

    a)   Level A
    b)   Level B
    c)   Level C
    d)   Level D

29) Several patients who were all riding on the same city bus, are being treated for similar symptoms: drooling, nasal discharge, vomiting, lacrimation and diarrhea. They also all have miosis. Which of the following is the most likely exposure to explain this?

    a)   exposure to a nerve agent
    b)   pepper spray
    c)   carbon monoxide
    d)   cyanide

30) Which of the following statements is most consistent with endocarditis?

    a)   Janeway lesions and Kernig's sign may be present.
    b)   Brudzinski's sign and Osler nodes may be present.
    c)   Roth's spots and Tinel's sign may be present.
    d)   Janeway lesions and Roth's spots may be present.

31) A patient has sustained chemical burns to his body while at work. What is the priority for this patient?

   a) Initiate two large-bore IVs and start an isotonic fluid bolus.
   b) Contact the poison control center for specific treatment guidelines.
   c) Remove the patient's clothes and irrigate the burn/exposed areas with copious amounts of fluid.
   d) Determine the specific chemical involved.

32) Which of the following sets of symptoms are most consistent with heat exhaustion?

   a) sweating, tachycardia, hypotension
   b) hot/dry skin, tachycardia, hypotension
   c) headache, nausea, dizziness
   d) core temp of 105.6

33) Which of the following is an appropriate treatment for acute external otitis?

   a) application of cool compresses
   b) performing a myringotomy
   c) insertion of an antibiotic soaked wick
   d) instructions for decongestant use

34) If untreated, rabies has approximately a _____ % fatality rate.

   a) 100
   b) 75
   c) 50
   d) 25

35) Which of the following body areas should not receive lidocaine with epinephrine for use as a local anesthetic?

a) the vermillion border
b) the scalp
c) the eyebrow
d) the pinna

36) Which of the following is the preferred patient position during the attempt to reduce a dislocated temporomandibular joint?

a) flat, supine
b) Trendelenburg
c) Fowler's
d) Reverse Trendelenburg

37) In the patient with a LeFort II fracture, free-floating movement would be noted in which of the following?

a) all facial bones
b) the teeth and lower maxilla
c) the nose and dental arch
d) unilateral periorbital area

38) A patient has been experiencing excessive weight gain, moon face, muscle wasting and truncal obesity. Which of the following most likely explains this?

a) Cushing's syndrome
b) Graves' disease
c) Syndrome of Inappropriate Antidiuretic Hormone (SIADH)
d) Addison's disease

39) A patient is exhibiting a fever of 104.9 and a rapid heart rate with a pulse of 154. Other history indicates they are taking levothyroxine (Synthroid). Which of the following should be suspected?

a) subacute thyroiditis
b) Graves' disease
c) myxedema coma
d) thyroid storm

40) A patient is complaining of abdominal cramping and persistent, foul smelling diarrhea. He reports recently that he has been on an extended outing in the Montana mountains for 3 weeks. Which of the following infections is most likely to be the cause?

a) amebiasis
b) giardiasis
c) amebic liver abscess
d) malaria

41) Which of the following is most likely to be found with a diagnosis of pyelonephritis?

a) low white blood cell (WBC) count
b) myoglobinuria
c) ketonuria
d) pyuria

42) A patient experiencing autonomic dysreflexia is having a bounding, severe headache. What other symptom is consistent with this condition?

a) hypotension
b) skin flushing below the level of the lesion
c) tachycardia
d) profuse sweating above the level of the lesion

43) Which of the following tests is most useful to guide treatment as part of a sexual assault evaluation?

a) pregnancy test
b) Rh factor
c) CBC
d) gonorrhea/chlamydia swabs

44) Which of the following vaginal infections does not require treatment of intimate sexual partners?

a) neisseria gonorrhea
b) chlamydia trachomatis
c) candida albicans
d) trichomonas vaginalis

45) Which of the following is of the highest priority in caring for a victim of sexual assault?

a) caring for any injuries sustained
b) evidence preservation
c) reporting of crime to appropriate law enforcement
d) ensuring that the family understands the situation

46) In the head injured patient requiring RSI (rapid sequence intubation), which of the following medications might be considered to be given before induction of paralysis with succinylcholine (Anectine), a neuromuscular blocker?

    a) Meperidine (Demerol)
    b) Lidocaine (Xylocaine)
    c) Ketamine (Ketalar)
    d) Atropine

47) A patient is brought by EMS with a chief complaint of lethargy. 1 day prior they were involved in a motor vehicle collision, hit their head on the windshield and refused care at that time. Since then they have been having headaches and drowsy. Their significant other reports they have been difficult to wake up. It is noted that the right pupil is fixed and dilated. The Glasgow Coma Scale score is 9. What is the most likely cause of these findings?

    a) postconcussion syndrome
    b) diffuse axonal injury
    c) epidural hematoma
    d) subdural hematoma

48) A patient is experiencing gingival pain, fever, chills, fatigue, bleeding from the gums and foul odor on the breath. The diagnosis is necrotizing ulcerative gingivitis. Which of the following is the common term for this condition?

    a) variant angina
    b) Prinzmetal's angina
    c) Ludwig's angina
    d) Vincent's angina

49) A 1 year old child has sustained a possible neck injury. Which of the following is a highly concerning finding?

   a) positive Babinski's reflex
   b) capillary refill of 3 seconds
   c) respirations of 30 per minute
   d) pulse of 60 per minute

50) Which of the following is true of Dilantin (phenytoin)?

   a) it potentiates the action of cardiac glycosides
   b) it is normally mixed in a D5W solution
   c) Rapid administration of may cause arrhythmias
   d) therapeutic levels are between 20-30 mg/ml

51) Treatment for myasthenia gravis (MG) is with anticholinesterase drugs. What is an antidote for anticholinesterase toxicity?

   a) Pyridostigmine (Mestinon)
   b) Physostigmine (Antilirium)
   c) Atropine
   d) Vitamin K (Aquamephyton)

52) Your patient has a suspected cervical spine injury and their airway is compromised. What should be done immediately?

   a) perform a head-tilt to open the airway
   b) perform a jaw-thrust, chin lift to open the airway
   c) perform blind orotracheal intubation with neck in midline position
   d) insert a nasopharyngeal airway

53) Autonomic dysreflexia is most associated with injuries at or above which level of the spinal cord?

a) T12
b) T6
c) L1
d) S5

54) Which of the following symptoms is most likely associated with a cluster headache?

a) epistaxis
b) aphasia
c) fever
d) lacrimation

55) Which of the following would likely be the earliest indicator of a change in a patient's neurological status?

a) change in level of consciousness (LOC)
b) delayed capillary refill
c) depressed motor response
d) sluggish pupillary responses

56) A patient who is 37 weeks gestation has fallen down a flight of stairs and is classified as a "trauma alert". Unless contraindicated, how should this patient be positioned?

a) knee to chest
b) left lateral recumbent
c) trendelenburg
d) reverse Trendelenburg

57) Which of the following indicates imminent delivery is most likely?

a) Braxton Hicks contractions
b) a bulging perineum
c) cervical dilation to 8cm
d) ruptured membranes

58) During delivery of the infant's head the emergency provider should:

a) apply fundal pressure
b) have mother push for a count of 10
c) have mother pant and apply gentle perineal pressure
d) tell mother to bear down

59) Which set of findings is most specific for bacterial conjunctivitis?

a) acute onset, mild pain, clear discharge
b) acute onset, moderate pain, purulent discharge
c) recurrent onset, no pain, no discharge
d) subacute onset, severe pain, preauricular lymphadenopathy

60) Which of the following is most likely associated with a retinal detachment?

a) vision decreasing and significant pain
b) gradual darkening of the visual field
c) flashing lights
d) intraocular pressure is increased

61) Your patient has sustained a blunt injury to the eye. Their complaints are blurred and blood-tinged vision in the affected eye. Your assessment shows blood in the anterior chamber of the eye. The most likely diagnosis is:

a) orbital fracture
b) globe rupture
c) retinal detachment
d) hyphema

62) Which of the following would most likely be associated with a penetrating eye injury?

a) redness and purulent drainage
b) increased intraocular pressure
c) normal visual acuity
d) irregular pupillary borders

63) A patient has an open wound to the mid femur area. There is deformity present however, distal neuro/vascular status is intact. Which of the following is the priority action at this point?

a) preparing for immediate wound closure
b) leaving the wound open to air
c) covering the wound with a moist, sterile dressing
d) irrigating the wound with a solution of betadine and hydrogen peroxide

64) Which of the following is most likely associated with an ankle dislocation?

a) muscle strain
b) bony fracture
c) tendon rupture
d) ligament sprain

65) Which of the following is seldom dislocated?

a) elbow
b) foot
c) shoulder
d) knee

66) A high school football player has received a significant blow to the right side, upper torso. He is presenting with his head tilted to the affected side, and his chin angled away from the injured side. Which of the following is most likely to have occurred?

a) humerus fracture
b) scapular fracture
c) clavicle fracture
d) shoulder fracture

67) When assessing for pulsus paradoxus, the clinician is considering what condition?

a) dissecting aortic aneurysm
b) tension pneumothorax
c) cardiac tamponade
d) elevated intracranial pressure

68) Your patient is an elderly female who has sustained a ground level fall. This patient is at highest risk for which fracture?

a) Cervical spine fractures
b) Pelvic fracture
c) Humerus fracture
d) Wrist fracture

69) Which of the following is an indication for splint application?

a) to prevent dislocations
b) to prevent damage to vascular and nerve supply
c) to align the bones of a comminuted fracture
d) to reduce a displaced fracture

70) Which of the following is a consideration to perform an emergent fasciotomy?

a) swelling of the extremity but pulseless
b) deformity of the lower leg and paralysis
c) a compartment pressure measurement of 19 mmHg
d) a hemophiliac with an expanding thigh hematoma and thigh deformity

71) The indications for intravenous thrombolytic (TpA) for pulmonary embolus include the documented presence of pulmonary embolus and which of the following?

a) an INR of 2.0
b) hypoxia
c) EKG changes
d) sudden loss of hemodynamic status

72) The patient is a 32 year old male who was a unrestrained passenger in a MVC. He is complaining of left thigh pain, neck pain and a headache. Vital signs per EMS on scene were: BP 110/68, HR 118, RR 24. EMS found his left thigh to be swollen and deformed and decreased pulse in the left dorsalis pedis. A sager traction device was placed in the field, 2 large bore IVs, 100% O2 via NRB mask, rigid cervical collar and long spine board. He has just arrived into the ED and been transferred to the trauma stretcher. The primary survey ABC's are intact and unchanged. During the secondary survey, you find the swelling is stabilized, and the distal pulse (dorsalis pedis) is equal to the unaffected leg. There is no paralysis or paresthesia in the distal left extremity. Which of the following is appropriate at this time regarding the traction splint?

a) change over to a Hare traction device
b) remove the splint as the neurovascular status is normal
c) leave the splint in place and continue with regular reassessments of the distal extremity
d) remove the splint and reasses the distal neurovascular status distally to see if the splint is still required.

73) Which nerve pathway is responsible for the condition and symptoms associated with carpal tunnel syndrome?

a) C7 nerve pathway
b) median nerve
c) ulnar nerve
d) radial nerve

74) An unresponsive infant in cardio-pulomnary arrest is being assessed. SIDS (Sudden Infant Death Syndrome) is suspected. Which of the following is most consistent with this differential?

    a) The child is described as usually lethargic, irritable and has been feeding poorly.
    b) The infant was recently healthy and found dead shortly after being put to sleep.
    c) The infant was considered ill and had several medical problems.
    d) The child had a history of being physically abused and was found unresponsive in the evening.

75) Under HIPAA regulation, which of the following are all considered to be sources of individually identifiable health information?

    a) name, diagnosis, date of birth
    b) date of birth, social security number, name
    c) diagnosis, social security number, date of birth
    d) allergies, date of birth, name

76) A 44 year old male who works as a professional, was unexpectedly laid off from his job 3 days ago. He is complaining of fatigue and lack of ability to cope. He states he's been drinking excessively during the last 72 hours. These findings are most likely associated with which of the following?

    a) depression
    b) situational crisis
    c) a manic episode
    d) alcoholism

77) Management of the potential organ donor includes which of the following?

   a) maintaining a CVP of no more than 2mmHg
   b) adjusting ventilator settings to maintain a PaO2 of 60mmHg
   c) maintaining the hematocrit (HCT) below 30%
   d) maintaining a urine output greater than 100 ml/hour

78) What is the most often reported emotional change following critical incident stress?

   a) depression
   b) hebephrenia
   c) euphoria
   d) panic

79) A patient showing dyspnea and shortness of breath is most often associated with which of the following?

   a) bradypnea
   b) tachycardia
   c) eupnea
   d) bradycardia

80) You are caring for a trauma patient. He has sustained a confirmed hemothorax. The following interventions and procedurs have been performed: chest tube placed, foley catheter placed, 2 large bore Ivs placed infusing isotonic fluids, and is being mechanically ventilated. Initially 450ml of blood immediately drained upon placement of the chest tube, and it continues to steadily drain. The patient's vitals after placement were: BP 144/72, HR 128 and RR at 16 (on ventilator). Which assessment parameter should be watched the closest over the next 60 min?

 a) urine output
 b) chest tube drainage
 c) central venous pressure
 d) blood pressure

81) A 30 year old male was an unrestrained driver in a MVC. He was ejected from the vehicle. Assessment reveals the following: he is conscious, BP 88/58, HR 132 and weak, RR 28 and shallow. Capillary refill is delayed. Lung sounds are present bilaterally, but diminished somewhat on the right side. Paradoxical chest wall movement is noted on the right side. Chest x-ray shows multiple rib fractures to ribs 4-7 on the right and a pneumothorax. The skin is pale, cool and moist. Which of the following best accounts for these findings?

 a) massive hemothorax
 b) ruptured diaphragm
 c) flail chest
 d) tension pneumothorax

82) Trauma care is best provided by which of the following?

   a) A university based trauma teaching fellowship.
   b) A research focused hospital with a progressive rehabilitation program.
   c) An integrated system from prevention and through rehabilitation.
   d) A level I trauma center.

83) The patient is a 40 y.o. firefighter who collapsed during an interior structure fire attack. He was "down" for 5 minutes before being rescued. He sustained inhalational injury from the heated gasses. He has been appropriately resuscitated and stabilized. He is intubated endotracheally and ventilated. Due to his injuries he is at risk for developing ARDS (Acute Respiratory Distress Syndrome). Which of the following best describes the main problem associated with ARDS?

   a) A decreased perfusion secondary to decreased cardiac contractility.
   b) An ineffective oxygenation due to bronchoconstriction and air trapping.
   c) A hypovolemia due to fluid third spacing.
   d) An impaired gas exchange due to lung non-compliance and altered pulmonary capillary permeability.

84) Peak expiratory flow rate (PEFR) is the most objective measurement of the patient's response to bronchodilator therapy. What is the optimal PEFR?

   a) PEFR variability of less than 30%.
   b) PEFR less than 50% of predicted or personal best.
   c) PEFR variability 20%-30%.
   d) PEFR greater than 80% of predicted or personal best.

85) Appropriate treatment for a rib fracture includes:

a) increasing fluids to prevent dehydration.
b) controlling pain to assist with breathing.
c) taping the chest wall to relieve pain/splint the affected area.
d) placing the patient in a supine position.

86) A 42 y.o. is complaining of a dry cough for 1 week. Now it has changed to a loose, productive cough. The vital signs are: HR 94, RR 18, T 99.0, BP 144/88. Which of the following is most likely?

a) COPD exacerbation
b) pneumonia
c) acute bronchitis
d) acute asthma exacerbation

87) A patient has sustained blunt chest trauma. Which of the following factors places this patient at highest risk?

a) a history of chronic intermittent asthma
b) a fracture of the 1st or 2nd rib
c) the patient is a child
d) there is bruising visible on the anterior chest wall

88) Which of the following is the priority intervention for a patient with a tracheal-bronchial rupture?

a) apply 100% O2 via non-rebreather
b) intubation with the cuff of the OETT placed beyond the rupture.
c) chest tube insertion
d) suctioning the airway

89) A 22 year old construction worker has sustained extensive burns to the head, face, chest and neck. Nearly all of the body hair from the waist up and his scalp hair has been singed or burned off due to the ignition of a flammable liquid. Which of the following would be most worrisome for significant complications for this patient?

a) a dry, intractable cough
b) a persistent wet cough
c) carbonaceous or black-tinged sputum
d) a carboxyhemoglobin of 3%

90) When considering crystalloid fluid resuscitation for hemorrhagic hypovolemic shock, which of the following is most appropriate?

a) infuse at 100-200 ml/kg
b) infuse at 80-90 ml/kg
c) infuse at 20-40 ml/kg
d) infuse at 5-10 ml/kg

91) A patient is being treated for shock. They have already received the appropriate amount of initial fluid resuscitation of normal saline and the most recent urine output has been calculated to be 35 ml/hr. Which would be the most appropriate next intervention?

a) Give another fluid bolus of 1 liter NS.
b) Give NS at 125 ml/hr and continue to monitor urine output.
c) Change IV solution to D5W.
d) Give another fluid bolus of 2 liters NS.

92) Early class I shock is best characterized by which of the following?

   a)   Increased systolic and diastolic pressures.
   b)   Falling systolic and diastolic pressures.
   c)   Rising systolic and falling diastolic pressures.
   d)   Normal or falling systolic and rising diastolic pressures.

93) Consider the following: BP 82/52, HR 132, RR 28/shallow, temp 102 F. Serum glucose 850 mg/dl, K+ 2.9 mEq/l, Na++ 130 mEq/L, Osmolality 365 mOsm/kg, Acetone negative. What would be the first intervention for this patient?

   a)   regular insulin
   b)   sodium bicarbonate
   c)   potassium chloride
   d)   normal saline IV

94) Which of the following statements best describes the findings in neurogenic shock?

   a)   a loss of parasympathetic vasomotor regulation
   b)   an inadequate cellular perfusion
   c)   a loss of sympathetic vasomotor regulation
   d)   a decrease in respiratory function

95) What is the priority intervention for a patient in hypovolemic shock?

   a)   insert a foley catheter to monitor renal/urine status
   b)   administer oxgyen
   c)   type and crossmatch
   d)   initiate IV access

96) Which of the following most accurately describes the pathophysiology of anaphylactic shock?

    a) decreased amounts of circulating catecholamines
    b) vascular endothelium damage
    c) systemic antigen-antibody response
    d) loss of sympathetic vasomotor function

97) What is the priority treatment for a patient in anaphylactic shock?

    a) administering an antihistamine
    b) maintaining airway patency
    c) administering epinephrine
    d) infusing a bolus of 500 ml NS

98) During a state of shock, what is the expected function/outcome of the renin-angiotensin apparatus?

    a) to increase urine output, to decrease reabsorption of Na+ and H20
    b) to increase urine output, to increase reabsorption of Na+ and H20
    c) to decrease urine output, to decrease reabsorption of Na+ and H20
    d) to decrease urine output, to increase reabsorption of Na+ and H20

99) While caring for a patient with suspected head injury, what is an appropriate priority intervention to help decrease the chance of the ICP increasing?

a) place in trendelenburg position.
b) hyperventilate the patient.
c) administer an isotonic crystalloid fluid bolus based on the patient's weight
d) perform a GCS (Glasgow coma scale) assessment.

100) Based solely on the information given, which of the following would be addressed during the secondary survey?

a) Absent lung sounds on the left.
b) Complete traumatic amputation below the left knee.
c) Inability to get chest rise and fall with bag-valve-mask ventilations.
d) Absent radial pulse, faint carotid pulse.

101) A 28 year old male has sustained an open femur fracture. Assessment shows: airway intact, 100% O2 via NRB mask is in place, distal pulses are present, there is no uncontrolled external hemorrhage, and GCS is 12. 2 liters of crystalloid IVF has been given as well as 2 units of PRBCs. Which of the following best indicates adequate volume replacement has been achieved at this time?

a) base excess +1
b) PaCO2 30 mmHg
c) blood pressure 86/50
d) HCO3 15

102) While considering which medication to use in treatment of cardiogenic shock, which of the following would be the desired beneficial outcome?

a) ↑ preload, ↑ peripheral resistance, ↑ afterload
b) ↓ cardiac output, ↓ cardiac contractility, ↓ peripheral resistance
c) ↓ peripheral resistance, ↓ afterload, ↑ cardiac output, ↑ contractility
d) ↓ preload, ↑ contractility, ↑ peripheral resistance

103) Which of the following are possible types of obstructive shock?

a) Cardiac tamponade, tension pneumothorax, pulmonary embolus
b) Aortic dissection, pulmonary embolus, congestive heart failure
c) Cardiac tamponade, massive acute MI, aortic dissection
d) Congestive heart failure, cardiac tamponade, coronary thrombus

104) All of the following trauma patients requires rapid blood replacement as part of their initial resuscitation. Which of the following would be the most appropriate to use O-NEG blood on?

a) 33 year old black male
b) 52 year old hispanic female
c) 24 year old caucasian male
d) 30 year old caucasian female

105) A dopamine infusion has been started at 200 mcg/minute on a 100 kg male trauma patient. What is the expected effect at this dose?

    a)   increased preload by reabsorption of Na+ & H2O
    b)   increased renal perfusion
    c)   increased afterload by less reabsorption of Na+ & H2O
    d)   decreased renal perfusion

106) Causes of acute renal failure can be generally classified into pre-renal, intra-renal or postrenal. In the emergency setting, prerenal is very common. Which of the following is a possible cause of prerenal failure?

    a)   renal tubular necrosis
    b)   an obstructing kidney stone
    c)   acute blood loss
    d)   renal artery stenosis

107) A trauma patient is being transported from one hospital to another. The initial lab panels showed a hematocrit of 39%. He was fluid resuscitated with 4 liters of crystalloid. He is intubated with the following settings: rate - 16, volume 500 ml, FiO2 - 40%. He has 2 large bore IV's and a foley catheter in place as well as a functioning chest tube that was placed to relieve a pneumothorax. On reassessment of basic labs, the hematocrit is noted to be 31%. Which of the following is the most likely cause for this decrease in hematocrit?

    a)   The specimen was hemolyzed.
    b)   The blood was drawn from above one of the IV sites.
    c)   The blood is hemodiluted due to IV fluid infusions.
    d)   Additional blood has been lost.

108) While treating a trauma patient with blunt thoracic trauma, the initial blood gasses are found to be: pH 7.20, PaCO2 44, PaO2 85, HCO3 15. Which of the following should be the next intervention based on those results?

    a)   administer 50 ml of 8.4% NaHCO3 IV
    b)   administer whole blood or PRBCs
    c)   start a bolus of LR at 40 ml/kg
    d)   maintain a liter of LR at keep open rate

109) What would you expect to see in a patient who is in hypovolemic shock and is on Metoprolol (Lopressor)?

    a)   bradycardia, hypertension & increased renin secretion
    b)   tachycardia, hypertension & decreased renin secretion
    c)   bradycardia, hypotension & decreased renin secretion
    d)   tachycardia, hypotension & increased renin secretion

110) Which of the following is predominantly a pre-load reducing agent?

    a)   Lopressor (metoprolol)
    b)   Norvasc (amlodipine)
    c)   Nitroglycerin
    d)   Hydralazine (apresoline)

111) Which of the following physical signs is associated with methanol ingestion?

    a) smell of garlic
    b) smell of formalin
    c) smell of mothballs
    d) smell of bitter almond

112) Bruising and/or ecchymosis/hematoma over the posterior flank area, possibly associated with renal injury or retroperitoneal hemorrhage is called:

a) Brown-sequard syndrome
b) Roth's spots
c) Gray-Turner's sign
d) Levine's sign

113) An elderly patient has the following blood gases: pH 7.24, HCO3 20, PaCO2 48. Which of the following toxicities is the most likely cause for this?

a) acetaminophen (Tylenol)
b) lanoxin (Digoxin)
c) aspirin (acetylsalicylic acid)
d) metoprolol (lopressor)

114) Over 3-4 hours, multiple adults have called 911 for evaluation for gastrointestinal symptoms. They are all experiencing nausea, vomiting and diarrhea with abdominal cramping. It is noted by the providers that they have all eaten at the same restaurant last evening. Based on this information you suspect which of the following?

a) salmonella
b) botulism
c) listeriosis
d) staphylococcus

115) Treatment of an overdose of a beta-adrenergic antagonist such as propranolol (Inderal) would be which of the following?

a) NaHCO3
b) Furosemide (Lasix)
c) Glucagon
d) CaCl

116) Which drug is used to treat an anticholinergic toxicity?

a) Atropine
b) Physostigmine
c) Lithium
d) Narcan

117) What medication would be indicated for tetany?

a) a paralytic
b) an opioid
c) an anticonvulsant
d) tetanus immune globulin

118) A camper returns from a 2 week camping trip in the NC mountains. Complaints are noted of malaise, abdominal pain, and myalgias. Exam shows a deep-red colored rash on the soles of the feet, the palms, wrist and ankles. Which of the following is the most likely cause for this presentation?

a) lyme disease
b) poison ivy
c) rocky mountain spotted fever
d) meningococcemia

119) Which is the preferred antimicrobial for the treatment of Rocky Mountain Spotted Fever?

a) Erythromycin
b) Bactrim
c) Tetracycline
d) Augmentin

120) Which of the following descriptions most accurately describes the brown recluse spider?

a) 2 circular black markings on the anterodorsal area.
b) Velvety black abdomen with brushes of red hair.
c) Hourglass shape ventrally.
d) Violin shaped mark antero-dorsally.

121) Which of the following has the highest rate of infection?

a) scorpion sting
b) human bite
c) dog bite
d) cat bite

122) A patient has sustained a laceration to the lower leg from a metal bar. He has never received tetanus immunization. Which of the following is most appropriate?

a) 0.5 ml tetanus toxoid and 250 units of tetanus immune globulin
b) 1 ml of tetanus toxoid
c) 250 units of tetanus immune globulin
d) 0.5 ml of tetanus toxoid

123) Which of the following is most specific for rabies?

a) increased salivation
b) depressed mental status
c) headache
d) hydrophobia

124) In a negligence case, what does the plaintiff have to prove?

a) lack of intent
b) mitigating circumstances
c) substandard delivery of care
d) intent to cause harm

125) Which piece of assessment data should be recorded in the record about every woman of childbearing age requesting medical care?

a) birth name
b) vaccination status
c) date of last menstrual period
d) number of pregnancies and live births

126) IV potassium is inadvertently administered to a patient while in ventricular tachycardia. As a result the patient winds up in ICU on a ventilator and suffers subsequent anoxic brain injury. The hospital settles financially with the family of the patient. Which of the following reflects the hospital's legitimacy of having the nurse repay the money lost?

a) vicarious liability
b) captain of the ship
c) indemnification
d) respondeat superior

127) Emergency pericardiocentesis has just been performed for cardiac tamponade. Which of the following best indicates the procedure was effective?

a) muffled heart sounds heard on auscultation
b) blood pressure is elevated
c) the patient is hypotensive
d) the aspirated fluid from the pericardium clots rapidly

128) An appropriate dose of tissue plasminogen activator (Tpa) has just been given to an acute STEMI patient. Which of the following is an outcome that suggests the treatment has been effective?

a) absence of arrhythmias
b) relief of chest pain
c) slight oozing of blood from the IV site
d) greater than 2mm elevation in the ST segment

129) A 34 year old male was kicked in the chest at a local rodeo show. His vital signs are as follows: BP 88/52, HR 138, RR 34. Physical exam reveals muffled heart sounds. A isotonic fluid bolus of 1 liter NS has been given. BP remains 86/50 and HR is 144. This patient most likely has which of the following?

a) aortic injury
b) cardiac tamponade
c) tension pneumothorax
d) myocardial contusion

130) What's the maximum number of times a patient may be defibrillated?

   a)   there is no limit
   b)   after a total dose of 10 joules/kg has been reached
   c)   10 times
   d)   20 times

131) What is the primary effect of vagal stimulation of the heart?

   a)   a decrease in sympathetic tone
   b)   a decrease in parasympathetic tone
   c)   an increase in sympathetic tone
   d)   an increase in parasympathetic tone

132) Occlusion of the left coronary artery (LCA) will most likely damage which area of the heart?

   a)   right ventricle
   b)   posterior myocardium
   c)   anterior myocardium
   d)   inferior myocardium

133) Which of the following is characteristic of a pericardial friction rub?

   a)   It is accentuated by having the patient lean forward.
   b)   It varies in intensity with respirations.
   c)   It is heard best at the apex using the bell of the stethoscope.
   d)   It is only heard during systole.

134) Which is the most common cause of cardiac arrest in adults?

   a) drug overdose/toxicity
   b) ventricular fibrillation
   c) respiratory failure/arrest
   d) electrolyte abnormalities

135) Which drugs would be most appropriate to treat a patient having acute pulmonary edema due to heart failure?

   a) Levophed, Inocor, Inderal
   b) Inocor, Digoxin, Dopamine
   c) Lasix, Morphine, Nitroglycerin
   d) Digoxin, Morphine, Lasix

136) The patient's history and exam reveal: fever and chills for 3 days, splinter hemorrhages of the nail beds and a systolic murmur. Which of the following is most likely?

   a) rheumatic endocarditis
   b) myocarditis
   c) endocarditis
   d) pericarditis

137) What is the best position to hear an S3 and S4 heart sound?

   a) right lateral
   b) high Fowler's
   c) supine
   d) left lateral

138) A patient suddenly goes into ventricular tachycardia. What should be done first?

a) administer lidocaine 1 mg/kg IV
b) assess the patient
c) defibrillate
d) derform a precordial thump

139) A patient has sustained massive internal injuries during a motor vehicle collision. The monitor shows sinus rhythm but there is no palpable pulse. Which of the following is the most appropriate intervention?

a) Begin dopamine IV titrate to blood pressure.
b) Perform a FAST.
c) Obtain a STAT chest xray.
d) Start infusing crystalloids and blood products.

140) Sudden onset of abdominal pain radiating to the flank in a 60 year old male with a history of hypertension, angina and congestive heart failure, mandates consideration for which of the following?

a) STEMI
b) abdominal aortic aneurysm
c) kidney stone
d) intermittent claudication

141) During administration of a thrombolytic agent (TpA) for an acute STEMI the patient reports the chest pain has decreased from #7 to a #2. You observe that he is also in an accelerated idioventricular rhythm and the blood pressure is 102/66. Respirations are 14/minute. Based on the above, which is the best answer?

   a) The patient is experiencing reperfusion symptoms.
   b) The patient is a candidate for an emergency pacemaker.
   c) The patient needs immediate transfer to a facility with cardio-thoracic surgery capability.
   d) The patient is having worsening ischemia.

142) Which of the following is one of the components of an emergency preparedness process?

   a) reassimilation
   b) intervention
   c) analysis
   d) mitigation & prevention

143) Which federal agency is responsible for US disaster management?

   a) Federal Bureau of Investigation (FBI)
   b) Disaster Medical Assistance Teams (DMATs)
   c) National Disaster Medical System (NDMS)
   d) Federal Emergency Management Agency (FEMA)

144) A patient with a core temperature of 90° F (32.2° C) would be expected to have which of the following?

a) cold, pale skin
b) rigid muscles, no shivering
c) apnea
d) shivering

145) A patient presents with complaints of left flank pain, urinary hesitancy and frank hematuria. A contrasted CT scan of the abdomen is ordered to evaluate for renal stone(s). Which of the following values indicates a potential concern before performing this radiographic study?

a) BUN = 15.0
b) Urine specific gravity = 1.015
c) Creatinine = 2.2
d) K+ = 4.8

146) Which of the following patients is the most appropriate to receive thrombolytics in a facility where cardiac catheterization is not an option?

a) 59 year old female with ST segment elevations in V1-V4, abdominal bruising and confusion, who was brought in after being involved in a MVC.
b) 56 year old male with chest pressure, diaphoresis, no ST segment elevations, and a new left bundle branch block (LBBB).
c) 52 year old female with HTN and diabetes, and showing a T wave inversion in II and III on her EKG.
d) 72 year old male with tachycardia, fever, SOB and ST depression in V5, V6 and I.

147) A patient has a 3 day history of sore throat, malaise, anorexia and fever. Inspection of the oro-pharynx shows a gray membrane covering the tonsils and posterior wall. Which of the following is most likely?

a) diphtheria
b) tonsillitis
c) haemophilus influenza type b (HiB)
d) pharyngitis

148) Which of the following statements reflects why alkali burns are more destructive than acid burns?

a) They cause major damage to fascia and muscle.
b) The produce coagulation necrosis.
c) They produce liquefaction necrosis.
d) They cause increased compartment pressures.

149) A 42 year old male collapsed while running a half-marathon. His core temp is 105.1 F (40.6 C). A urine sample is obtained and it is tea-colored. Based on the scenario, what is the most likely cause for this?

a) The urine is concentrated in the hyperthermic patient.
b) There are red blood cells in his urine.
c) Myoglobin is present in the urine.
d) The patient is dehydrated.

150) Which type of learning takes the longest to achieve?

a) affective
b) social
c) psychomotor
d) cognitive

Part 2: Answers & Rationales

1) ANSWER: b) a distended bladder

A distended bladder is an absolute contraindication to peritoneal lavage. An indwelling urinary catheter must be placed first to decompress the bladder. There is no contrast dye used during this so an allergy to it isn't a factor. A history of abdominal surgery is not related to the indication for this procedure. An unconscious patient is not a contraindication, actually peritoneal lavage may be useful in the unconscious patient as they are unable to give a history or report their symptoms.

2) ANSWER: b) diverticulitis

Approximately 50% of adults age 60-80 have diverticulitis. Inflammatory bowel disease is more likely in the 10-30 year old age range. Bowel obstructions are not likely related to intussusception. They are more likely related to surgical scar tissue, adhesions, tumors or fecal impactions. Ulcerative colitis is more likely in the 30-50 age range.

3) ANSWER: a) bowel sounds in the chest.

Bowel sounds in the chest are more likely associated with a diaphragmatic rupture. Blood in naso-gastric contents, absent bowel sounds and epigastric tenderness are all more likely associated with a stomach injury.

4)   ANSWER: d)   Magnesium citrate

Vasopressin would be inappropriate via an NG tube. It is used IV during resuscitative situations. Heparin would not be appropriate or safe without knowing this patient's risk for bleeding due to their liver disease. Additionally, heparin is not given via the NG route. PTU could be given via an NG tube, but is is used to treat thyroid storm/thyroid crisis. The question does not indicate this patient is experiencing this condition. Magnesium citrate can be given via the PO/NG route. In this liver failure patient, this would be a possible choice as the action of Mag Citrate to rid the bowel of blood and fecal matter would also serve to reduce the ammonia levels, which in a liver disease patient, may be elevated already.

5)   ANSWER: a)   Murphy's sign

Murphy's sign is associated with acute choleycystitis. It is elicited when the right upper quadrant is palpated. The acute tenderness elicited is significant enough to cause the patient to "catch their breath", as well as report significant pain. Murphy's sign is best elicited by "hooking" the fingers up and under the rib cage in the direction of the liver. Psoas sign is associated with possible appendicitis. It is demonstrated with increased pain in the right lower quadrant with hyperextension and elevation of the right hip. Cullen's sign is associated with intra-abdominal hemorrhage. It is seen as bruising/ecchymosis around the umbilicus. Lhermitte's sign is an electric like shock pain that shoots throughout the body produced by neck flexion. It is associated with some conditions such as traumatic cervical spine injury or multiple sclerosis.

6)   ANSWER: c)   alcohol abuse and biliary tract disease

Alcoholism and biliary tract disease account for nearly 80% of all cases of pancreatitis.

7) ANSWER: a) pancreatitis

While there are possible fluid volume deficits associated with all of these conditions, pancreatitis is associated with them most often and most significantly.

8) ANSWER: d) abdominal distention and bloody diarrhea

Commonly associated findings of mesenteric infarction are vomiting, abdominal pain, abdominal tenderness, abdominal distention, bloody diarrhea and hypotension. Constipation is more likely associated with an intestinal obstruction as well as vomiting fecal-like material. Abdominal distention and free air are associated with a perforated viscous.

9) ANSWER: c) pericardial friction rub

The pericardial effusion associated with a pericarditis is attributable to a pericardial friction rub. In pericarditis the lungs are usually clear to auscultation. The chest pain of pericarditis is usually relieved by leaning forward. The presence of an S3 heart sound should alert to the presence of left sided heart failure.

10) ANSWER: a) seizures

Lidocaine is used in cardiovascular emergencies as an antiarrhythmic however it has effects in the CNS as well. The likelihood of toxic effects is dose dependent. Early symptoms of lidocaine toxicity may be as subtle as tinnitus or a metallic or tingling sensation in the mouth. Seizures (generally grand mal type) are witnessed at higher toxic levels. Cushings response is associated with intra-cranial pressure increases. Tachycardia and hypertension would not be expected as part of lidocaine toxicity.

11) ANSWER: b) Deceleration forces and shearing

All of the above answer choices "may" cause aortic trauma but the most specific primary injury pattern associated with aortic trauma is the deceleration forces which cause a shearing of the aorta and rupture. Penetrating and blunt injuries as well can cause aortic damage but they are not as specific.

12) ANSWER: b) pulse amplitude greater in the arms than the legs

With a ruptured descending aorta the pulse amplitude is greater in the upper extremities than the lower due to the decreased perfusion to the lower extremities. A difference in left versus right upper extremity blood pressure suggests a subclavian artery rupture not aortic. Distended jugulars and muffled heart sounds are suggestive of cardiac tamponade (Beck's triad).

13) ANSWER: d) acute myocardial infarction

The pain of an acute myocardial infarction is the LEAST likely of the choices to be relieved by nitroglycerin. In an AMI, the vessel is so occluded that the coronary vasodilation is not as likely to be enough to restore flow. Nitroglycerin more consistently and reliably works in angina pectoris as the occlusion is not complete or total, and the resultant coronary dilation is more likely to allow for enough blood flow to ease the anginal pain. Heart failure isn't commonly associated with substernal chest pain. Pericarditis shouldn't respond to nitroglycerin as the problem is not in the coronary arteries, but the pericardial sac.

14) ANSWER: d) hypothermia

Recall the ACLS mnemonic "H's and T's" to cover these. Hypothermia is one of the reversible causes. Even though digitalis toxicity is a "toxin" ("T" for t-ablets, or t-oxins), consider the mechanism of toxicity for digoxin. An overdose of digoxin would lead to considerable slowing of AV node conduction and resulting in a decreased heart rate, blocks and eventually asystole or PEA. While ventricular fibrillation is a possibility with digoxin toxicity, it is not as specific as hypothermia.

15) ANSWER: a) to cause a temporary asystole

The purpose behind defibrillation is to completely depolarize the myocardium which should then allow the heart's natural intrinsic pacemaker to take over and generate an organized rhythm. Defibrillation does not synchronize the rhythm nor will it "jump start" the heart. It may or may not produce a sinus rhythm with successful defibrillation.

16) ANSWER: a) verapamil

Giving verapamil to a patient with v-tach can be lethal. It may accelerate the heart rate and decrease blood pressure, more so in patients with atrial fibrillation and WPW syndrome. The other options may be indicated for wide complex tachycardia depending on the scenario.

17) ANSWER: c) to reduce myocardial oxygen consumption

Beta-adrenergic blockers (beta-blockers: metoprolol etc) reduce myocardial oxygen consumption and demands of the ischemic areas of the myocardium. They reduce infarct size by decreasing sympathetic tone and reduction of afterload.

18) ANSWER: c) bradycardia, peaked T waves and a wide QRS

Hyperkalemia may present as peaked T waves, a widened QRS and bradycardia as well as the disappearance of P waves and eventually idioventricular rhythm and asystole.

19) ANSWER: c) Anterior wall of the heart

The left main coronary artery feeds both the anterior and lateral wall of the myocardium. The right coronary artery (RCA) usually supplies the inferior wall. A branch of the posterior descending artery supplies the posterior wall. The right ventricle is supplied by the RCA as well.

20) ANSWER: b) Streptokinase (Streptase)

Streptokinase has the highest incidence of associated allergic reactions, approximately 5% of the time. Current practice is to co-administer steroids and antihistamines before it is administered.

21) ANSWER: a) afterload

Afterload is the force that the ventricle must overcome during ejection or systole. Systemic vascular resistance is the resistance to this. Preload is the amount of stretch on the myocardium before systole. Stroke volume is the amount of blood ejected by the left ventricle.

22) ANSWER: b) S3 or ventricular gallop

Dilated cardiomyopathy (heart failure) and post-myocardial infarction are two common causes of an S3 (third heart sound) or ventricular gallop. A systolic and diastolic mumur are both non-specific as there are many possible causes for these. The S1 sound is a normal heart sound.

23) ANSWER: d) medulla

The medulla controls the arterioles, blood pressure and the rate and depth of respirations. Significant injury to the medulla usually is fatal. The frontal lobe controls personality, judgement, thought and logic. The diencephalon contains the thalmus - the sensory pathway between the spinal cord and the brain cortex. The hypothalmus regulates body temperature, heart rate, appetite and sleep.

24) ANSWER: c) apply c-spine immobilization

This is a good example of a professional board level national exam question. In this case it is asking for an INTERVENTION. The only two answers that are interventions are "c" and "d". "d" requires a physician's order, it is a dependent action. "c" is an independent action and the only intervention acceptable at this point in the scenario. "a" and "b" are assessments.

25) ANSWER: d) paresthesias of the extremities

Exopthalmos is the bulging of the eyeballs, commonly associated with hyperthyroidism. Contamination of CSF is not specific to a spinal cord injury. In neurogenic shock, secondary to a spinal cord injury, one would expect hypotension and bradycardia. Paresthesias of the extremities would be most likely associated with a spinal cord injury due to damage of the transmitting neurons/nerve pathways.

26) ANSWER: c) decrebrate posturing

Decerebrate or decorticate posturing are both highly specific to an increased intracranial pressure (and most likely - near herniation). Hypertension is part of Cushing's response, however hypertension alone is not highly specific for one pathology. Likewise with bradycardia - although it is part of Cushing's response it is not specific. Unequal pupils may be associated with increasing ICP but, also, there are many other potential causes as well.

27) ANSWER: a) to decrease preload

Diuretics decrease preload by decreasing the circulating blood volume, (they also exert a veno-dilation effect) thereby reducing the filling pressure of the right side of the heart (preload). They may affect the afterload but only secondary to a decreased ejection by the left side of the heart. Diuretics should have no effect on cardiac contractility. Diuretics may decrease myocardial oxygen demand but this is not their purpose.

28) ANSWER: c) Level C

Level D is a standard work uniform with no respiratory protection. Levels A and B require the user to be completely encapsulated in a suit and isn't necessary when dealing with the patient. Levels A and B are specialty suits, use is usually limited to professional haz-mat workers. Level C is a tyvec suit, gloves and purified positive air respirator. It is easy to learn, can be worn for sustained time and the air supply isn't limited, and is appropriate for health care workers.

29) ANSWER: a)   exposure to a nerve agent

Pepper spray would cause skin irritation and burning. Miosis would not be likely. Carbon monoxide poisoning is usually asymptomatic, except for terminal findings of cherry red mucous membranes or lips. Cyanide poisoning would present with seizures, altered mental status and the smell of bitter almonds. Nerve agents (organophosphates) cause a cholinergic crisis systemically with findings of too much acetylcholine present. The symptoms of a cholinergic crisis may be summarized by the memory key words of "MUDDLES" or "SLUDGEM".

30) ANSWER: d)   Janeway lesions and Roth's spots may be present.

Kernig's and Brudzinski's signs are elicited when testing for meningeal irritation (meningitis or subarachnoid bleed). Tinel's sign is elicited when testing for possible carpal tunnel syndrome. Osler nodes are painful nodes found on the hands and feet, they are associated with possible endocarditis. Janeway lesions are painless, reddened, nodules found on the hands and feet. They are highly pathognomonic for endocarditis. Roth's spots are retinal hemorrhages which are strongly associated with endocarditis.

31) ANSWER: c)   Remove the patient's clothes and irrigate the burn/exposed areas with copious amounts of fluid.

Decontamination and irrigation are the priority. Potential further damage to the patient needs to be stopped as well as to ensure safety for the care team. Copious amounts of water will be used. There is no need to contact poison control at this point as decontamination is the priority. IV fluids will be important but not as the priority intervention. Determining the chemical involved will be important, however, decontamination first is more important.

32)   ANSWER: c)   headache, nausea, dizziness

Hypotension is more consistent with heat stroke than heat exhaustion. A core temp of 105.6 is consistent with heat stroke as well. The remaining answer of headache, nausea, dizziness - while not overtly ill, is the only plausible answer considering hypotension and a core temp of 105.6 are NOT consistent with heat exhaustion.

33)   ANSWER: c)   insertion of an antibiotic soaked wick

Topical antibiotics via a soaked wick inserted into the external auditory canal are indicated. These would be preferred versus PO antibiotics. Decongestant usage and myringotomy would be indicated for otitis media, not externa. Compresses should be hot/warm for external otitis.

34)   ANSWER: a)   100

Once symptoms of the disease begin to manifest, it is nearly always fatal. Http://www.who.int/mediacentre/factsheets/fs099/en

35)   ANSWER: d)   the pinna

Epinephrine should not be used on the "fingers, nose, penis, ears, toes". These areas lack good peripheral circulation and epinephrine has a potent vaso-constricting effect. The vaso-constriction may be enough to cause localized tissue ischemia and necrosis and should be avoided in these areas. The other areas mentioned are safe to use epinephrine in.

36) ANSWER: c)   Fowler's

For reducing a TMJ dislocation the provider should be above the patient and the patient in Fowler's position with good back support. Constant downward pressure on the mandible will facilitate it to slide backward into the proper alignment. The other positions listed don't facilitate this downward and backward movement.

37) ANSWER: c)   the nose and dental arch

The LeFort II fracture is a triangular type fracture which involves the central portion of the maxilla across the upper nasal area. It may also extend into the orbit. The result is a free-floating nasal and dental arch area/bones. A free floating periorbital area is not associated with this pathology. LeFort I fractures are associated with free floating teeth and maxillary area. Lefort III fractures involve all the facial bones being free floating.

38) ANSWER: a)   Cushing's syndrome

The symptoms described are common findings with Cushing's syndrome. Cushing's syndrome is a disorder with increased levels of glucocorticoids and corticotropin. Addison's disease would exhibit hyperpigmentation, changes in sexual characteristics, and dehydration. SIADH would exhibit weight gain, nausea, behavioral changes and anorexia. Graves' disease is a result of hyperthyroidism and evidenced by exophthalmos, tremors and goiter.

39) ANSWER: d)   thyroid storm

Thyroid storm is hyperthyroidism exaggerated by stress or infection. Fever and tachycardia are typical manifestations. Other symptoms may include hypotension, vomiting, hyperreflexia and irritability. Graves' disease manifests as hyperthyroidism, goiter and exopthalmos. Myxedema coma is a severe form of hypothyroidism and is typified by coma, hypothermia and hyponatremia. Subacute thyroiditis is a self-limiting inflammation of the thyroid gland usually from a viral infection.

40) ANSWER: b)   giardiasis

Giardiasis is a water borne parasitic infection contracted by the ingestion of untreated or inadequately treated water. Malaria doesn't typically have diarrhea, but it commonly has nausea, fever, chills and sweating. Amebiasis is an infection of the large intestine contracted by ingestion of food or water contaminated by feces. The symptoms of Amebiasis are chronic versus acute and the diarrhea alternates with constipation. Amebic liver abscess is an extraintestinal complication of amebiasis. The symptoms don't typically include diarrhea.

41) ANSWER: d)   pyuria

Pyelonephritis would include leukocytosis (elevated WBCs), hematuria, pyuria and bacteriuria. The patient may also have back/flank pain, fever, chills and/or nausea/vomiting. Ketonuria is more likely associated with a diabetic state or a state of dehydration. Myoglobinuria is indicative of muscle wasting or breakdown such as in rhabdomyolysis.

42) ANSWER: d) profuse sweating above the level of the lesion

Acute autonomic dysreflexia is a reaction of the autonomic (involuntary) nervous system to overstimulation. It is characterized by paroxysmal hypertension (the sudden onset of severe high blood pressure) associated with throbbing headaches, profuse sweating, nasal stuffiness, flushing of the skin above the level of the lesion, slow heart rate, anxiety, and sometimes by cognitive impairment. The sympathetic discharge that occurs is usually in association with spinal cord injury (SCI) or diseases such as multiple sclerosis.

43) ANSWER: a) pregnancy test

A pregnancy test is indicated because if positive, pregnancy prevention medication (Plan B) cannot be administered. Rh factor test is only pertinent to the pregnant female experiencing vaginal bleeding. CBC is a nonspecific test. GC/Chlamydia swabs would not be routinely available for a couple of days. Prophylaxis with antibiotics would be started empirically regardless.

44) ANSWER: c) candida albicans

Candida is a yeast infection and not considered communicable. Gonorrhea, chlamydia and trichomonas are all sexually transmitted/transmittable infections and as such notification and treatment of sexual partners is required.

45) ANSWER: a) caring for any injuries sustained

The patient's health and safety always takes precedence. While evidence preservation is very important, it is secondary to assessing and treating their health and any injuries. Reporting of sexual assault is not universally mandatory. Concern for the family is important, but it must be only with the patient's permission to share her health/medical information.

46) ANSWER: b) Lidocaine (Xylocaine)

Succinylcholine may increase ICP (intracranial pressure) in the patient with head injuries. Giving 1mg/kg of lidocaine prior to administration of succinylcholine provides ICP control. Atropine would be considered in pediatric scenarios so as to blunt a vagal response and the resultant bradycardia. Ketamine and Demerol are contraindicated in head injuries as they may increase ICP.

47) ANSWER: d) subdural hematoma

The signs and symptoms of a subdural hematoma (SDH) may not appear for up to 48 hours due to the slower progression of symptoms. It is typically a venous bleed and more subtle in onset. Classic symptoms of an epidural hematoma (EDH) include a transient initial loss of consciousness followed by a return to baseline, and then shortly thereafter a rapid decline and possibly fatal. The time period between return to baseline and decline is called the "lucid interval". A diffuse axonal injury is typified with an acute sudden presentation, a loss of consciousness and posturing. Post concussion syndrome presents as a history of head injury with prolonged, recurrent symptoms related to the initial injury. Common symptoms are: dizziness, irritability, insomnia, judgement impairment, memory loss, tinnitus and others. It is not considered fatal and should have no lateralizing symptoms.

48) ANSWER: d) Vincent's angina

Necrotizing ulcerative gingivitis is also known as Vincent's angina. Ludwig's angina is a cellulitis, soft tissue infection of the submandibular tissue and the floor of the mouth and possibly extending into the neck. The concern with Ludwig's angina is the potential for involvement of the soft tissues involving the airway and possible airway obstruction. Vincent's angina, is called "trench mouth". Prinzmetal and variant angina refer to coronary artery spasm (non atherosclerotic).

49) ANSWER: d) pulse of 60 per minute

The normal heart rate for a 1 year old is 90-120. Bradycardia is a sign of increasing ICP or an oxygenation problem. Respirations for a 1 year old range from 20-30. Capillary refill of up to 2-3 seconds is acceptable for a 1 year old. A positive Babinski's reflex is acceptable under age 2 or prior to walking.

50) ANSWER: c) rapid administration may cause arrhythmias

Phenytoin should not be given faster than 50 mg/minute. If given faster than that it may suppress the myocardium and lead to arrhythmias and cardiac arrest. Therapeutic levels should be between 10-20 mg/ml. Phenytoin can only be mixed with normal saline. Phenytoin inhibits the action of cardiac glycosides and steroids.

51)   ANSWER: c)   Atropine

A person having an anticholinesterase toxicity is experiencing a high level of acetylcholine, and in a "cholinergic" crisis. Atropine is a powerful anti-cholinergic that would begin to reverse that. Vitamin K is the antidote/treatment for coumadin toxicity/overdose. Physostigmine and Pyridostigmine are anticholinesterase medications used to treat myasthenia. They could both be causes of an anticholinesterase toxicity.

52)   ANSWER: b)   perform a jaw-thrust, chin lift to open the airway

The preferred method for opening the airway when suspecting a cervical spine injury is the jaw thrust, chin lift maneuver. This provides the optimal protection to the cervical spine. Inserting a nasopharyngeal airway is useful to support an open airway but not as the primary means of opening it. Head-tilt maneuver should only be used in the patient NOT suspected of cervical spine injury. Blind orotracheal intubation should only be attempted after basic airway maneuvers have been ineffective; possibly as a back up airway as well.

53)   ANSWER: b)   T6

Autonomic dysreflexia is most commonly associated with injuries at and above T6. It is a serious hypertensive condition triggered by a noxious stimuli which triggers an exaggerated sympathetic nervous system response resulting in a hypertensive emergency.

54) ANSWER: d) lacrimation

Cluster headaches are typically one sided, periorbital area, and occur in clusters or groups and recur. May also be associated with lacrimation, tearing or nasal congestion. Epistaxis is usually not associated with headaches. Fever is a non-specific finding and usually cluster headaches do not have fever. Aphasia would be associated with a possible stroke presentation.

55) ANSWER: a) change in level of consciousness (LOC)

The earliest indicator of a change in neuro status is the LOC. Any patient with a change in their LOC should be re-evaluated for other changes as well. A comprehensive neuro assessment is always key, but the LOC is an absolute. Capillary refill indicates tissue perfusion not neuro status. Motor responses and pupillary reactions are indicative of neuro status but they are later signs, later than changes in LOC.

56) ANSWER: b) left lateral recumbent

This position avoids direct compression of the inferior vena cava due to the weight of the gravid uterus while the patient is lying supine. Up to 30% of the venous return may be compromised and decreasing the preload to the heart. Trendelenburgs (and reverse Trendelenburg) would cause vena cava compression. Knee to chest position is used during management of a prolapsed cord during delivery and not indicated in trauma management.

57) ANSWER: b)   a bulging perineum

A bulging perineum indicates the head is far down into the vaginal canal and delivery is imminent. Membranes may rupture early on in the labor progression not necessarily when it's imminent. Cervical dilation to 10 cm needs to occur before the patient begins pushing. Braxton Hicks contractions are "false labor".

58) ANSWER: c)   have mother pant and apply gentle perineal pressure

The risk of perineal tears are high if the mother pushes during the moment of delivery. Having her pant prevents her from bearing down/pushing/valsalva. Gentle perineal pressure helps to prevent an explosive delivery. Fundal pressure during delivery isn't necessary.

59) ANSWER: b)   acute onset, moderate pain, purulent discharge

Bacterial conjunctivitis has an acute onset and purulent (pus) discharge/drainage. The pain may be mild to moderate. Viral conjunctivitis would have acute onset and clear discharge. Recurrent symptoms is more typical with allergic conjunctivitis.

60) ANSWER: c)   flashing lights

Symptoms of a retinal detachment may include: painless decrease in vision, cloudy or smoky vision, flashing lights. They may also describe vision decreasing like a veil or curtain being pulled. (Not the same as darkening). The intraocular pressure will be normal or low.

61) ANSWER: d) hyphema

The most frequent symptoms of a hyphema include impaired visual acuity, blod in the anterior chamber, and blood tinged vision. Commonly precipitated by some blunt injury. Globe rupture would present with decreased acuity, irregular pupillary border and decreased intraocular pressure. A retinal detachment would likely have smoky or cloudy vision, flashing lights or floaters and a "veil" or "curtain" coming down over the visual field. An orbital fracture would affect the extraocular movements, possible lip paresthesias and possibly a sunken eyeball (enopthalmos).

62) ANSWER: d) irregular pupillary borders

Redness and drainage are most likely associated with conjunctivitis. Increasing intraocular pressure (above 20 mmHg) is indicative of Glaucoma most often. The patient with a penetrating eye injury would have alterations in visual acuity most likely. Irregular pupillary borders (sometimes called a "teardrop" pupil) are most often seen with this.

63) ANSWER: c) covering the wound with a moist, sterile dressing

An open wound near the site of a suspected fracture (this patient has a suspected femur fracture) is considered an open fracture until proven otherwise. At this point, keeping it free from subsequent contamination with a sterile, moist dressing is appropriate. Irrigation and wound closure will be left until the determination is made if it is a fracture or not.

64) ANSWER: b) bony fracture

Ankle dislocations have a high rate of concurrent bony fractures of the distal tibia or fibula. The others do not happen as common as the fractures.

65) ANSWER: b) foot

Dislocations of the foot are very rare (not to be confused with the ankle). Dislocations of the knee, shoulder and elbow are much more common.

66) ANSWER: c) clavicle fracture

These symptoms are characteristic of a fractured clavicle. Additionally the patient cannot usually raise the arm on the injured side. A shoulder fracture is actually a shoulder dislocation with associated bony fracture. Scapular fractures are uncommon and considered a high velocity/high force injury pattern. The strong muscles of the chest/thoracic wall typically hold this fracture in place. The head would not be tilted or leaning in this case.

67) ANSWER: c) cardiac tamponade

Typical symptoms of cardiac tamponade (whether a medical pericarditis or traumatic hemorrhage) are referred to as Beck's Triad (hypotension, JVD, and muffled heart sounds). Pulsus paradoxus looks for a change in the systolic blood pressure of 10 mmHg or more on inspiration vs exhalation. This finding is nearly pathognomonic for cardiac tamponade.

68) ANSWER: b) pelvic fracture

Pelvic and lower extremity fractures are the most common fall-associated fractures in the elderly. Humerus and cervical spine fractures aren't necessarily age related. Wrist fractures are most often associated with a fall on an outstretched hand.

69) ANSWER: b) to prevent damage to vascular and nerve supply

The first and foremost guiding principle of medicine/nursing is to prevent further harm/do no harm. In this case application of a splint will stabilize the fracture site, lessening the likelihood of subsequent damage to the neurovascular supply to the distal extremity. Splinting may or may not prevent dislocations, it may or may not re-align bones (but not likely with a comminuted fracture), and it may or may not reduce a displaced fracture: but this is not the primary purpose of splinting.

70) ANSWER: b) deformity of the lower leg and paralysis

An emergent fasciotomy is considered with significant tissue trauma/crush injuries. As the pressure increases in compartment syndrome the last pathways to become compromised are the motor and sensory pathways. The arterial pulse will be lost before this. Therefore the presence of either paralysis or numbness indicates the compartment pressure is at a critical level. The maximum physiological compartment pressure is 20mmHg or less. A pressure of 19mmHg doesn't mandate emergent fasciotomy. Simply a hematoma and deformity isn't as high a risk for limb compromise as deformity, possible fracture and paralysis.

71) ANSWER: d) sudden loss of hemodynamic status

A patient with an INR of 2.0 would not be a good candidate as they are already anticoagulated. Hypoxia, even with a documented PE, should be still treated with supplemental oxygen first. EKG changes associated with a pulmonary embolism are often the S1-Q3-T3 pattern. These are just findings, they are not compromising hemodynamics. S1-Q3-T3 doesn't always show up and isn't 100% specific for a PE. In the patient with a known pulmonary embolus (one that has already been documented by imaging/scan), who then loses or drops their hemodynamics/blood pressure - the use of TpA may be considered. This is from the FDA approval labeling.

72) ANSWER: c) leave the splint in place and continue with regular reassessments of the distal extremity

There is no need to change from a Sager to a Hare traction device. If applied correctly, they both accomplish the same purpose. Changing devices may compromise the stability of the fracture site and the effect of the tamponade effect in the thigh/femur compartment. Removing the splint may be warranted under certain circumstances, however, just because the neurovascular status is normal is not a reason to remove the splint. Removing the splint and reassessing for changes also is not warranted. This action may be considered if radiographs were negative for fracture - the question stem does not indicate that. Leaving the splint in place and doing regular reassessments is the most prudent course of action at this time.

73) ANSWER: b) median nerve

Symptoms of CTS are intermittent numbness of the thumb, index, and middle fingers and the radial side of the ring finger. This region is innervated by the median nerve. Numbness and paresthesias in the median nerve distribution are the hallmark. The numbness often occurs at night. It can be relieved by wearing a wrist splint that prevents flexion. Pain in carpal tunnel syndrome is primarily numbness that is so intense that it wakes one from sleep. Some of the individual predisposing factors include: diabetes, obesity, pregnancy, and hypothyroidism. Occupational causes involve use of the hand and arm, such as heavy manual work, work with vibrating tools, and highly repetitive tasks even if they involve low force motions.

74) ANSWER: b) The infant was recently healthy and found dead shortly after being put to sleep.

While SIDS has no specific diagnostic criteria, it does have some factors more commonly associated with it than not. Most often, SIDS occurs in an otherwise healthy child and are usually found shortly after being put down to bed. There is usually no history of poor intake, lethargy or significant comorbidities. Physical abuse may increase the chances for SIDS but it is not conclusively proven to be a causation.

75) ANSWER: b) date of birth, social security number, name

Under HIPAA, these are all considered pieces of individually identifiable health information. Diagnosis and allergies are not.

76)   ANSWER: b) situational crisis

A situational crisis is the result of a specific event in one's life. They are overwhelmed by it and react emotionally. Fatigue, insomnia, and lack of adequate decision making capacity are common. The situational crisis may go as far as to precipitate behaviors which, in turn, cause their own crisis (IE: alcohol or drug abuse/mis-use). There is not enough information given to label this patient as an alcoholic. Mania would demonstrate euphoria, a labile affect and hyperactivity. Depression symptoms are usually present for more than 2 weeks and more gradual in onset.

77)   ANSWER: d) maintaining a urine output greater than 100 ml/hour

A CVP of 2mmHg (or less), a PaO2 of 60mmHg and a HCT <30% are all too low for normal physiological parameters. These parameters would inadequately be supporting the organism. A steady, high rate of urine output ensures adequate renal functioning and is a good indicator or organ/tissue perfusion.

78)   ANSWER: d)   panic

Panic occurs most commonly in response to post-critical incident stress. Typically euphoria, hebephrenia and depression aren't associated with critical incident stress and are associated with other problems/disorders.

79) ANSWER: b) tachycardia

Dyspnea and shortness of breath cause a sympathetic nervous system stimulation with increased amounts of circulating catecholamines which causes tachycardia. Bradycardia would be more likely to be seen with vagal stimulation. Eupnea is a normal respiratory rate. Bradypnea is a low respiratory rate, this would not be an expected finding in respiratory distress or shortness of breath.

80) ANSWER: b) chest tube drainage

The initial drainage was a significant amount which indicates the internal bleeding was significant. If the drainage continues at a significant rate (some authors quote 200 ml/hr) then the chest needs to be surgically opened and explored. Urine output, vital signs and CVP are all important, but the amount of chest drainage has the most serious impact/implication for this patient.

81) ANSWER: c) flail chest

Flail chest is defined as 2 or more adjacent ribs with fractures in 2 or more places. The resulting "free floating" segment gives rise to the paradoxical chest wall movement which is specific for a flail chest. Flail chest injuries have a high mortality due to the difficulty in effective ventilation as well as the significant underlying tissue damage. A hemothorax would have diminished or decreased lung sounds, fluid present on chest x-ray and dullness to percussion over the blood collection area. Bowel sounds into the chest cavity are pathognomonic for ruptured diaphragm. Tension pneumothorax would have absent or markedly decreased/diminished lung sounds, respiratory distress, JVD and tracheal deviation (late sign). Paradoxical chest wall movement is nearly pathognomonic for flail chest injury.

82) ANSWER: c) An integrated system from prevention and through rehabilitation.

The best trauma care is provided in a comprehensive trauma service that encompasses the full spectrum of trauma care from prevention/education, through point of injury, initial care/resuscitation, rehabilitation and community re-integration. Not all level 1 trauma centers provide the "best" care if they are also lacking in any of the aforementioned disciplines. The American College of Surgeons has recommended trauma care across the entire continuum as the standard by which others are measured. This is best accomplished when prevention through rehabilitation are part of that spectrum.

83) ANSWER: d) An impaired gas exchange due to lung non-compliance and altered pulmonary capillary permeability.

ARDS is typically secondary to a systemic pulmonary insult, in this case the inhalational injury. In the acute phase the pulmonary capillary permeability increases as protein rich fluid rushes in to the affected lung tissues as part of the inflammatory response. Subsequently there is a rapid onset of pulmonary edema which quickly decreases the functional lung space. What follows is hypoxia from intrapulmonary shunting, decreased lung capacity and the presence of pulmonary edema. Initially the cardiac output/contractility is not affected, but may be seen as a consequence in later stages. Bronchoconstriction and air trapping are more consistent with COPD presentations. There may be a hypovolemia develop, but it is not the primary complication in ARDS.

84) ANSWER: d) PEFR greater than 80% of predicted or personal best.

The optimal PEFR is greater than 80% of predicted or personal best with a variability of less than 20%. The PEFR indicates the response of the airways to bronchodilator therapy when administered for exacerbations of asthma or COPD, or anyone with a reactive airway or bronchoconstriction problem. As the patient responds beneficially to the treatment(s) the PEFR will increase, thus demonstrating increasing bronchodilation and the ability to overcome the obstructive nature of these attacks. A decreasing PEFR with increasing symptoms indicates further medical management or adjustment is necessary.

85) ANSWER: b) controlling pain to assist with breathing.

Pain control for the patient with a rib fracture is paramount. This will allow adequate expansion of the lung tissue to prevent atelectasis and subsequent pneumonia. Other measures of proven benefit are turning, coughing, deep breathing and incentive spirometry. The patient should be placed in high fowler's position to aid in gas exchange and breathing effort. Taping of the chest wall promotes/leads to atelectasis and possible pneumonia. A common occurrence with rib fractures is an associated pulmonary contusion - so fluids should be judiciously administered to prevent pulmonary edema.

86) ANSWER: c)   acute bronchitis

Typical symptoms for bronchitis are a cough and productive of phlegm. Fever is not typical with bronchitis and a normal respiratory rate is more common. COPD exacerbation would more likely have only a dry cough, shortness of breath, and an elevated respiratory rate. Pneumonia would likely have a fever and an elevated respiratory rate and a heart rate over 100. Acute asthma exacerbation would not typically have a cough.

87) ANSWER: b)   a fracture of the 1st or 2nd rib

A fracture of the 1st or 2nd rib is considered very significant. It requires substantial force and trauma to do this. The risk to the underlying tissue/structures is very significant as well: risk for pneumothorax, subclavian artery/vein injury and neck structure injuries. A 1st/2nd rib fracture requires vigilance and a thorough assessment for any concurrent injuries. Chronic intermittent asthma is typically managed in the outpatient setting and is considered a stable condition. While a child presents special considerations and a different approach to trauma, by itself, just being a child is not the highest risk presented here. Visible bruising on the chest wall may be associated with other underlying injuries, but by itself, it does not warrant the highest concern compared to a 1st or 2nd rib fracture.

88) ANSWER: d)   suctioning the airway

While the definitive answer for treatment of a tracheal-bronchial rupture is to pass the endotracheal tube's cuff, beyond/past the point of rupture, it is not the priority. Following the priorities of patient assessment and care (A,B,C,D), the highest priority is "A" for airway. In this case, suctioning the airway is the priority. Next would be "B" for breathing which would include the 100% NRB mask and intubation. Chest tube insertion may be considered during the "B" step if it was indicated.

89) ANSWER: c) carbonaceous or black-tinged sputum

A dry, intractable cough and persistent wet cough both may be associated with a possible inhalational injury, but they are not specific, nor as significant at this time. Indeed the presence of both would necessitate frequent re-assessments for progression of symptoms and a change in the patient's clinical course. A carboxyhemoglobin of 3% is the top end of normal (normal being approximately 1-3%). The carbonaceous or black-tinged sputum is very specific for pulmonary/inhalation injury. The presence of such indicates the significance of inhaled (possibly superheated) gases as well as foreign materials in the fumes. This patient mandates vigilance in reassessment and monitoring due to the likelihood of airway/pulmonary tissue reactive edema and possible compromise. It is likely the provider would consider early intubation to secure and protect this potentially compromised airway and pulmonary tree.

90) ANSWER: c) infuse at 20-40 ml/kg

Generally accepted practice for IV fluid resuscitation for an unstable patient in hypovolemic/hemorrhagic shock is rapid infusion of a crystalloid at 20-40 ml/kg. The minimal rate of 5-10 ml/kg may not be enough to adequately raise the circulating volume. A higher rate of 80-200 ml/kg may result in a significant dilution of RBCs, platelets and coagulation factors. Additionally clot formation may be disrupted at the sites of injured tissue.

91) ANSWER: b) Give NS at 125 ml/hr and continue to monitor urine output.

Based on the urine output (> 30 ml/hr), this indicates adequate renal perfusion and as such the IV fluids should be decreased and urine output continued to be monitored. There is no information given that indicates the need to switch to D5W at this time.

92) ANSWER: d) Normal or falling systolic and rising diastolic pressures.

The chemoreceptors in the carotid and aorta sense a decrease in oxygenation and a rising carbon dioxide. This causes a release of catecholamines which induce peripheral vasoconstriction and increase the peripheral resistance. The net result is an increase in diastolic pressure as a means to increase preload and thus cardiac output. The systolic pressure is gradually falling due to the loss of circulating volume.

93) ANSWER: d) normal saline IV

Based on the information given, this patient is most likely in HHNK. The high glucose, the negative acetone, the mildly lowered K+ and the osmolality show this. This patient is suffering from an osmotic diuresis secondary to the hyperglycemia. They are in a state of profound dehydration and need fluids. They will need insulin and possibly potassium as well, but they need fluids FIRST. Sodium bicarbonate is not necessary as they are not in an acidotic state.

94) ANSWER: c)   a loss of sympathetic vasomotor regulation

Neurogenic shock results from insult/damage to the spinal cord above the 6th thoracic level. This insult affets the autonomic nervous system and the sympathetic outflow. Due to this there is a block of any sympathetic vasomotor control and regulation below the level of the injury. The resultant parasympatheic affect is unopposed causing venous dilation, decrewased preload, decreased cardic output. The unopposed parasympathetic stimulation causes a vagal response and a subsequent bradycardia.

95) ANSWER: b)   administer oxgyen

Type and crossmatch is not an intervention, it is a diagnostic or assessment modality. Of the others (insert catheter, giving oxygen, starting IV access), giving oxygen occurs first when following the priorities of care: A-airway, B-breathing, C-circulation.

96) ANSWER: c)   systemic antigen-antibody response

The profound, systemic vascular dilation is caused by the dilation effects of histamine. This results in a "hi-space" shock; distributive. Neurogenic shock is characterized by the loss of sympathetic vasomotor function. Septic shock results from vascular endothelial damage due to bacterial endotoxins. In shock like states, circulating catecholamines are circulating in increased amounts as part of the stress response.

97) ANSWER: b) maintaining airway patency

Following the paradigm of Primary Survey (ABCDE) then Secondary Survey, airway comes first regardless of the problem. Epinephrine would be next to increase systemic vascular tone and bronchodilate, probably simultaneously with an antihistamine to address the cause. IV fluid bolus may be helpful, but would not be definitive nor would it occur before stabilizing the airway.

98) ANSWER: d) to decrease urine output, to increase reabsorption of Na+ and H20

In a shock state the renin-angiotensin system works to conserve/increase the circulating volume. Thus increasing perfusion to tissues and increasing preload.

99) ANSWER: b) hyperventilate the patient

Positioning the patient in REVERSE trendelenburg might be an option (head elevated), but not trendelenburg (feet elevated) as that would increase ICP. Administering an isotonic fluid bolus would increase the circulating volume and increase the preload/CVP, this would increase the ICP. Performing a GCS assessment is an assessment not an intervention. The question asks for an intervention. Hyperventilation (to reduce the PaCO2) might be of use as a reduced circulating PaCO2 causes a degree of vasoconstriction, thereby, reducing the ICP as the vasculature would occupy less space. However, hyperventilation must be used with caution because at a certain point it can cause arterial constriction also, effectively reducing the amount of oxygenated blood being carried to the insulted brain tissue when it needs it the most.

100) ANSWER: b)   Complete traumatic amputation below the left knee.

Without further description to indicate there is life threatening (IE: exsanguinating) bleeding from the amputation site - this is not a life threatening problem and would be dealt with during the secondary survey. Absent lung sounds indicate a tension pneumothorax which is a life threat. Inability to get chest rise and fall with ventilations is a life threat. Absent radial pulse and faint carotid indicate a systolic blood pressure of 60 mmHg at best, this is hypoperfusion and needs to be dealt with during the primary survey during the "C" circulation step (start IV access and fluid bolus). Not all amputations are life threatening and the description here does not indicate that.

101) ANSWER: a)   base excess +1

Normal PaCO2 runs 35-45 mmHg. A PaCO2 of 30 mmHg indicates the patient is loosing CO2 possibly in a compensation attempt for a metabolic acidosis such as lactic acidosis associated with inadequate cellular metabolism/resuscitation. Blood pressure of 86/50 yields a MAP of 62. Current recommendations are to resuscitate to a minimum MAP of 65. HCO3 of 15 is too low which indicates the pt may be in an acidotic state consistent with a lactic acidosis, and poor perfusion. Base excess of +1 is within the normal range of -2 to +2 (some authors go with -3 to +3).

102) ANSWER: c) ↓ peripheral resistance, ↓ afterload, ↑ cardiac output, ↑ contractility

In cardiogenic shock the heart itself is not functioning at it's capacity and the output is reduced. The ideal medication to improve the cardiac output and perfusion would DECREASE peripheral resistance, which reduces the amount of force necessary for the heart to contact and eject blood. Also DECREASING afterload additionally reduces the workload of the heart as well as it's oxygen demand. INCREASING cardiac output is the primary goal. INCREASING contractility improves the amount of blood ejected each systole, thereby increasing the cardiac output.

103) ANSWER: a) Cardiac tamponade, tension pneumothorax, pulmonary embolus

Obstructive shock is described as a process that impairs the blood flow from the heart and great vessels resulting in hypotension and decreased tissue perfusion. Congestive heart failure and a massive acute MI would be types of cardiogenic shock. A coronary thrombus would impede blood flow to the myocardium causing an acute MI.

104) ANSWER: d) 30 year old caucasian female

Type O-neg blood is available in only a small percentage of blood donors in the US. Because of this it should be used only in those specific populations that it would be the safest choice. In the case of the female of childbearing status, administering type O-neg avoids the potential for the Rh iso-immunization which may then complicate future pregnancies if there is a maternal-fetal incompatibility. (Or the current, undiagnosed pregnancy). While there is no set age for childbearing years, the 30 year old is more likely to be childbearing potentially than the 52 year old.

105)　ANSWER:　b) increased renal perfusion

The infusion in the question is running at 2 mcg/kg/min. At this dose dopamine predominantly exerts an effect on the kidneys to increase their perfusion. This is called a "renal dose" for dopamine. At higher levels (generally above 5 mcg/kg/min) results in increased peripheral vasoconstriction, increased afterload and elevation of the blood pressure.

106)　ANSWER:　c)　acute blood loss

Prerenal kidney failure is typically a problem with getting fluid or the circulating volume to the kidneys. The kidneys require fluid present to function appropriately. In the emergency setting, acute renal failure is most commonly prerenal in nature and common causes are acute blood loss and dehydration.

107)　ANSWER:　c)　The blood is hemodiluted due to IV fluid infusions.

Hematocrit is not affected by hemolysis. Additional blood may have been lost, but more likely due to the massive crystalloid replacement, this is the cause. If the blood was drawn from above one of the IV sites, there possibly could be a hemodilutional effect, but not likely to this magnitude.

108) ANSWER: b) administer whole blood or PRBCs

This is a trauma patient and the ABG's reveal a metabolic acidosis. Based on this information we must assume they are acidotic due to inadequate tissue perfusion. While administering NaHCO3 would address the acidosis, it doesn't address the problem. In this case, the best intervention to address the problem is to increase the oxygen carrying capacity of the circulating volume, and as such, to administer blood. A bolus of LR only increases the circulating volume, it doesn't increase the oxygenation to the tissues. A keep open rate would be useless.

109) ANSWER: c) bradycardia, hypotension & decreased renin secretion

The normal response to a shock-like state involves activation of the sympathetic nervous system causing increased heart rate, blood pressure and renin secretion. A patient taking a beta-blocker such as metoprolol (Lopressor) would exhibit the opposite of these responses (bradycardia, hypotension and decreased renin secretion). An otherwise healthy patient NOT on a beta blocker would exhibit tachycardia, hypotension and increased renin secretion.

110) ANSWER: c) Nitroglycerin

Nitroglycerin, in normal therapeutic doses, predominately works to reduce preload. At higher doses it is an afterload reducing agent as well. Lopresor (metoprolol) predominantly works on the beta receptors of the heart. A reduction in blood pressure is a consequence of the reduction in heart rate. Norvasc (amlodipine) and hydralazine (apresoline) both work predominately on afterload.

111) ANSWER: b) smell of formalin

Formalin is a distinctive odor associated with methanol poisoning/exposure. Formic acid is a byproduct of methanol. Bitter almonds is associated with cyanide toxicity. Moth balls odor is typical of camphor and naphthalene. Garlic is typical of arsenic and organophosphate exposures.

112) ANSWER: c)   Gray-Turner's sign

Gray-Turner's sign is a physical finding suggestive of possible retroperitoneal hemorrhage and/or kidney damage. Brown-sequard syndrome is a set of symptoms associated with a specific spinal cord hemisection injury, it does not cause an outward skin coloration finding. Roth's spots are small hemorrhages seen on the retina associated with micro-emboli suggestive of endocarditis. Levine's sign is when a patient having chest pain puts their clenched fist over their sternum, suggestive of acute coronary syndrome or acute myocardial ischemia.

113) ANSWER: c)   aspirin (acetylsalicylic acid)

These ABGs reflect a metabolic acidosis. Aspirin toxidromes cause a gap (metabolic) acidosis. The elderly are more prone to this due to decreased renal functioning. Another hallmark symptom of ASA toxidrome is tinnitus. Tylenol toxidromes cause liver damage initially, later a metabolic acidosis may ensue, but not initially. Digoxin toxicity would result in bradycardia, heart blocks and asystole. Lopressor toxicity would also result in a slowing of the cardiac conduction and thus the heart rate and subsequent blood pressure.

114) ANSWER: a)   salmonella

Salmonella symptoms typically present within 12-24 hours of the ingestion. Foods more likely to carry salmonella include: milk, egg dishes, salad dressings, shellfish and poultry. Staphylococcal infections usually appear much more suddenly, typically within 1-6 hours and usually have headache and fever. Listeriosis occurs usually 3-21 days after exposure/ingestion. Diarrhea is not a usual finding in Botulism.

115) ANSWER: c)   Glucagon

In a beta-blocker overdose such as propranolol (Inderal), the adverse effect is a bradycardia and possible asystole. In this case glucagon is the agent of choice to treat this. It acts as an inotrope to increase the heart rate circumventing the blocked beta-1 receptors in the heart. NaHCO3 is used to treat the acidosis associated commonly with salicylate and tricyclic antidepressant overdoses. Furosemide is not a specific antidote for any toxidromes. Calcium chloride would be used for a calcium channel blocker overdose.

116) ANSWER: b)   Physostigmine

Physostigmine is an acetylcholinesterase inhibitor given in anticholinergic syndromes. It reverses the anticholinergic effects (delirium, somnolence) caused by certain anticholinergics. Atropine is an anticholinergic. It would cause anticholinergic toxicity. Atropine is the antidote of choice for organophosphate toxicity/exposure. Lithium is not an antidote for any condition. Narcan is an antidote for an opiate overdose.

117) ANSWER: d) tetanus immune globulin

Two possible causes of muscle tetany are hypocalcemia and tetanus infection (Clostridium tetani). In this case, the only possible treatment from the options listed is to administer tetanus immune globulin. Additional treatment for C. tetani would include administration of tetanus toxoid and penicillin or tetracycline. A paralytic would only paralyze the patient to facilitate possible airway intervention but would not mitigate the true effects of the pathogen. Opioids may be useful to ease the pain of the muscle spasm, but this is not definitive therapy. Seizures aren't associated with C. tetani and anticonvulsants wouldn't be expected to be administered.

118) ANSWER: c) rocky mountain spotted fever

The highest incidences of RMSF are found in NC, SC, OK and VA. Symptoms typically present within two weeks of contact with infected ticks. A red colored rash starts on the soles, palms, hands, feet, wrists and ankles. It will become petechial and spread to the rest of the body. Poison ivy is typically linear streaks and accompanied by burning, pruritus and blistering. Meningococcemia typically has an upper respiratory prodrome and will follow with fevers, chills, rigors, delirium and a purpuric rash, as well as with nuchal rigidity. Lyme disease typically has the rash starting within 1 week. There is one lesion at the location of the tick bite and is described as a "bull's eye" lesion or rash.

119) ANSWER: c) Tetracycline

Tetracycline (25-50 mg/kg/day) is the preferred, effective agent in treatment of RMSF. All the others are not effective against it.

120) ANSWER: d)   Violin shaped mark antero-dorsally.

A velvety black abdomen with brushes of red hair are found on the funnel web spider. An hourglass shape ventrally is the black widow. The circular black markings on the anterodorsal area are more consistent with fruit flies.

121) ANSWER: d)   cat bite

20-50% of cat bites will become infected. 10-50% of human bites become infected (most of these will be hands). 3-20% of dog bites become infected. Other than the routine risk of cellulitis, there is no evidence that scorpion bites become infected. While there is a large amount of potentially pathological pathogens in human bite wounds, typically they don't become infected unless the hands are involved. Additionally human bite wounds are concerning for potential for HIV transmission, although rare.

122) ANSWER: a)   0.5 ml tetanus toxoid and 250 units of tetanus immune globulin

In the patient who has never received tetanus immunization (no primary series), tetanus immune globulin is indicated. This patient needs protection today with immune globulin. This patient also will receive tetanus toxoid to begin the long term protection via active immunity.

123) ANSWER: d)   hydrophobia

Hydrophobia is pathognomonic for rabies infection. This is due to painful spasms that occur in the upper airway and hypopharynx, especially when swallowing. The other symptoms are possible, but they are not as specific to the rabies infection as hydrophobia.

124) ANSWER: c)   substandard delivery of care

Negligence does not require proof of intent or lack of intent to cause harm. Mitigating circumstances would be brought up by the defendant not the plaintiff. Negligence is an unintentional tort or a civil wrong done without intent to the plaintiff.

125) ANSWER: c)   date of last menstrual period

The date of the last menstrual period will affect medications, imaging and procedures. The other information is important and good to know, but none of them carry as much decision making impact as the potential for pregnancy.

126) ANSWER: c)   indemnification

If an employee is found liable for negligence, then vicarious liability requires the employer to pay a settlement. Following that, the employer has the option to require indemnification from the employee for the losses the employer incurred. Respondeat superior is the vicarious liability the employer has for negligent acts of employees who are acting within the scope of their employment. Captain of the ship refers to any negligence being attributable to the physician due to their role as "captain of the ship".

127) ANSWER: b)   blood pressure is elevated

Cardiac tamponade is an obstructive type shock process and hypotension is one of the findings. Performance of emergency pericardiocentesis should result in improving the blood pressure as the constriction around the heart is relieved and the heart can achieve increased filling and subsequent increased ejection. Pericardial blood doesn't clot. A successful pericardiocentesis would negate muffled heart sounds.

128) ANSWER: b) relief of chest pain

Expected findings associated with effective Tpa administration include: relief of chest pain, reperfusion arrhythmias, and normalization of the ST segment. Oozing of blood from IV sites may occur after Tpa administration, but this does not indicate successful STEMI treatment.

129) ANSWER: b) cardiac tamponade

The muffled heart sounds on physical exam plus the mechanism of injury suggest a cardiac tamponade. Cardiac tamponade may not respond to a fluid bolus/challenge as the patient is not suffering from hypovolemia. A myocardial contusion is possible from the mechanism of injury, however, it's usual findings are anterior/inferior EKG changes and would not have muffled heart sounds. A tension pneumothorax would present with absent or markedly diminished/decreased lung sounds on the affected side. Heart sounds would be normal. An aortic injury would have a wide mediastinum on chest x-ray, but should not have muffled heart sounds.

130) ANSWER: a) Tthere is no limit

There is no maximum number of times a patient may be defibrillated, however, if they remain in ventricular fibrillation or pulseless ventricular tachycardia for more than 30 minutes - the chances of survival are low.

131) ANSWER: d) an increase in parasympathetic tone

Vagal stimulation increases parasympathetic tone which decreases the heart rate and may slow AV conduction.

132) ANSWER: c)   anterior myocardium

The LCA supplies the anterior and lateral myocardium. The inferior myocardium and right ventricle are usually supplied by the right coronary artery (RCA). The posterior myocardium is supplied by the posterior descending artery (PDA).

133) ANSWER: a)   It is accentuated by having the patient lean forward.

A pericardial friction rub may be heard when the pericardial surfaces are inflamed. It is indicative of pericarditis. It is audible over the entire pericardium and best heard with the diaphragm of the stethoscope. A pericardial friction rub does not vary with respirations, that would be a pleural rub. Pericardial rubs are heard both during systole and diastole and are increased in intensity with the patient leaning forward.

134) ANSWER: b)   ventricular fibrillation

Ventricular fibrillation is the most common cause of cardiac arrest in adults. The others may all lead to cardiac arrest, however they are not as common as ventricular fibrillation.

135) ANSWER: c)   Lasix, Morphine, Nitroglycerin

Drug treatment for acute pulmonary edema is directed at preload reduction. Lasix, morphine and nitroglycerin are all first line preload reducing agents. Digoxin is an inotrope and in the setting of acute heart failure it may increase cardiac workload and it doesn't address the preload problem. Inocor may help with cardiac output, but it's main role is not as a preload reduction agent. Dopamine and levophed would increase the cardiac workload.

136) ANSWER: c)   endocarditis

These findings are classic for endocarditis. Additional findings associated with endocarditis are Janeway lesions on the fingertips and Roth's spots on the retina. Rheumatic endocarditis usually shows symptoms of left sided heart failure, SOB, crackles and/or wheezes. Myocarditis usually has fatigue, dyspnea, palpitations and an S3 heart sound and systolic murmur. The classic symptoms of pericarditis are chest pain relieved by leaning forward and a precordial friction rub may be heard of during auscultation.

137) ANSWER: d)   left lateral

S3 and S4 are best heard with the bell of the stethoscope placed over the point of maximal impulse with the patient in the left lateral position.

138) ANSWER: b)   assess the patient

Assessment always precedes intervention. In this case it is essential to determine if the patient is hemodynamically stable or unstable as this will dictate further interventions.

139) ANSWER: d)   Start infusing crystalloids and blood products.

In this scenario, the most likely cause of the patient's PEA is massive volume loss. ("H" for hypovolemia). Dopamine may increase the blood pressure but at the expense of worsening perfusion. Performing a FAST and/or chest xray might be useful to assess for other pathologies such as cardiac tamponade or hemothorax, however they are assessments and not interventions.

140) ANSWER: b)   abdominal aortic aneurysm

Abdominal pain with a sudden onset must lend itself to consideration for an abdominal aortic aneurysm. The kidney stone is a possibility but not the same priority. These symptoms are not typical of a STEMI. Intermittent claudication is the symptoms associated with peripheral arterial disease, akin to angina in the patient with coronary artery disease.

141) ANSWER: a)   The patient is experiencing reperfusion symptoms.

Resolving chest pain and an accelerated idioventricular rhythm may be associated with reperfusion or resolving ischemia following TpA administration. Worsening ischemia would be more likely with increasing chest pain or the chest pain unchanged. The idioventricular rhythm associated with reperfusion is usually transient and does not require emergency pacemaker as the patient is perfusing. There is no indication for cardio-thoracic surgery at this time.

142) ANSWER: d)   mitigation & prevention

Components of an emergency preparedness program include: Mitigation & Prevention, Planning, Response, Recovery.

143) ANSWER: d) Federal Emergency Management Agency (FEMA)

FEMA is responsible for disaster management at all governmental levels. FEMA works in many different types of disasters (natural, medical, man-made etc). NDMS assists in the treatment and movement of patients in disaster responses/disaster scenes. It also provides a national hospital network to accept patients during a national emergency or disaster. DMATs are part of NDMS. They are specialized medical teams providing the hands-on patient care. The FBI is an investigational agency, performing criminal investigations on a federal level.

144) ANSWER: d) shivering

A core temperature of 90° F (32.2° C) is considered mild hypothermia. Shivering does not start to diminish until below 90° F (32.2° C). Skin will still be normal skin tone in mild hypothermia. Apnea would occur with severe/profound hypothermia, usually < 82.4° F (28° C).

145) ANSWER: c) Creatinine = 2.2

A creatinine of 2.2 is too high and indicates the presence of renal disease/insult. An ensuing contrast load administered would further insult and adversely affect the kidneys. This patient should not receive IV contrast. A bun of 15 is normal as well as a specific gravity of 1.015 and a K+ of 4.8.

146) ANSWER: b)   56 year old male with chest pressure, diaphoresis, no ST segment elevations, and a new left bundle branch block (LBBB).

The presence of a new LBBB is one criteria for an acute myocardial infarction. This patient needs reperfusion therapy. A cardiac catheterization is preferred, however in its absence, thrombolysis should be started if no contraindications are present. Patient "A" has findings of an acute MI, however, thrombolysis is contraindicated in the setting of trauma/bleeding as evidenced by the abdominal bruising. Patient "C" is not having an acute MI and thrombolysis is not warranted. Patient "D" has ST depression which indicates a cardiac pathology potentially, but not an acute MI. His symptoms also suggest a possible alternate pathology.

147) ANSWER: a)   diphtheria

The usual presentation for diphtheria is the formation within two to three days of onset a thick coating can build up in the throat making it very hard to breathe and swallow. This thick gray coating is called a "pseudomembrane". The pseudomembrane is formed from dead tissue caused by the toxin that is produced by the bacteria. Tonsillitis would just be inflamed/injected and swollen tonsils. Pharyngitis would be inflammation/injection of the posterior pharynx and possibly the tonsils as well. H. flu infection of the upper respiratory tree is non-specific in it's presentation.

148) ANSWER: c)   They produce liquefaction necrosis.

Alkali burns cause liquefaction necrosis, deeper damage, less pain and less tissue tension - all of which increase the likelihood to spread. Treatment must include copious irrigation for a minimum of 30-40 minutes. Acid burns cause coagulation necrosis. Fascia and muscle damage is more likely with electrical burns. Alkali burns do not typically cause compartment syndrome.

149) ANSWER: c)   Myoglobin is present in the urine.

Tea-colored urine is a descriptive word to indicate the presence of myoglobin. Myoglobin is released when muscle cells break down. Strenuous/exertional activities are a common cause of this. The urine may be concentrated if the patient is dehydrated, but the tea-colored descriptor is not specific for that. Red blood cells in the urine indicate a different pathology than the potential for rhabdo. They are not as specific as myoglobin is.

150) ANSWER: a)   affective

Affective learning includes incorporating though one's attitudes, beliefs and values. Changes to this mode require a lengthy time to integrate and evolve with a new learned behavior pattern. Cognitive learning is usually the quickest as it doesn't require a change in attitudes or belief systems. Psychomotor follows cognitive learning. It requires a skill set to be learned which means practice and muscle memory. Social learning, which is learning by observation of others, may or may not require belief or value change and thus it isn't guaranteed to be as long a process as affective learning, which does require manipulation to values and beliefs.

Part 3: Notes

1) Ⓐ Ⓑ Ⓒ Ⓓ	41) Ⓐ Ⓑ Ⓒ Ⓓ	81) Ⓐ Ⓑ Ⓒ Ⓓ	121) Ⓐ Ⓑ Ⓒ Ⓓ
2) Ⓐ Ⓑ Ⓒ Ⓓ	42) Ⓐ Ⓑ Ⓒ Ⓓ	82) Ⓐ Ⓑ Ⓒ Ⓓ	122) Ⓐ Ⓑ Ⓒ Ⓓ
3) Ⓐ Ⓑ Ⓒ Ⓓ	43) Ⓐ Ⓑ Ⓒ Ⓓ	83) Ⓐ Ⓑ Ⓒ Ⓓ	123) Ⓐ Ⓑ Ⓒ Ⓓ
4) Ⓐ Ⓑ Ⓒ Ⓓ	44) Ⓐ Ⓑ Ⓒ Ⓓ	84) Ⓐ Ⓑ Ⓒ Ⓓ	124) Ⓐ Ⓑ Ⓒ Ⓓ
5) Ⓐ Ⓑ Ⓒ Ⓓ	45) Ⓐ Ⓑ Ⓒ Ⓓ	85) Ⓐ Ⓑ Ⓒ Ⓓ	125) Ⓐ Ⓑ Ⓒ Ⓓ
6) Ⓐ Ⓑ Ⓒ Ⓓ	46) Ⓐ Ⓑ Ⓒ Ⓓ	86) Ⓐ Ⓑ Ⓒ Ⓓ	126) Ⓐ Ⓑ Ⓒ Ⓓ
7) Ⓐ Ⓑ Ⓒ Ⓓ	47) Ⓐ Ⓑ Ⓒ Ⓓ	87) Ⓐ Ⓑ Ⓒ Ⓓ	127) Ⓐ Ⓑ Ⓒ Ⓓ
8) Ⓐ Ⓑ Ⓒ Ⓓ	48) Ⓐ Ⓑ Ⓒ Ⓓ	88) Ⓐ Ⓑ Ⓒ Ⓓ	128) Ⓐ Ⓑ Ⓒ Ⓓ
9) Ⓐ Ⓑ Ⓒ Ⓓ	49) Ⓐ Ⓑ Ⓒ Ⓓ	89) Ⓐ Ⓑ Ⓒ Ⓓ	129) Ⓐ Ⓑ Ⓒ Ⓓ
10) Ⓐ Ⓑ Ⓒ Ⓓ	50) Ⓐ Ⓑ Ⓒ Ⓓ	90) Ⓐ Ⓑ Ⓒ Ⓓ	130) Ⓐ Ⓑ Ⓒ Ⓓ
11) Ⓐ Ⓑ Ⓒ Ⓓ	51) Ⓐ Ⓑ Ⓒ Ⓓ	91) Ⓐ Ⓑ Ⓒ Ⓓ	131) Ⓐ Ⓑ Ⓒ Ⓓ
12) Ⓐ Ⓑ Ⓒ Ⓓ	52) Ⓐ Ⓑ Ⓒ Ⓓ	92) Ⓐ Ⓑ Ⓒ Ⓓ	132) Ⓐ Ⓑ Ⓒ Ⓓ
13) Ⓐ Ⓑ Ⓒ Ⓓ	53) Ⓐ Ⓑ Ⓒ Ⓓ	93) Ⓐ Ⓑ Ⓒ Ⓓ	133) Ⓐ Ⓑ Ⓒ Ⓓ
14) Ⓐ Ⓑ Ⓒ Ⓓ	54) Ⓐ Ⓑ Ⓒ Ⓓ	94) Ⓐ Ⓑ Ⓒ Ⓓ	134) Ⓐ Ⓑ Ⓒ Ⓓ
15) Ⓐ Ⓑ Ⓒ Ⓓ	55) Ⓐ Ⓑ Ⓒ Ⓓ	95) Ⓐ Ⓑ Ⓒ Ⓓ	135) Ⓐ Ⓑ Ⓒ Ⓓ
16) Ⓐ Ⓑ Ⓒ Ⓓ	56) Ⓐ Ⓑ Ⓒ Ⓓ	96) Ⓐ Ⓑ Ⓒ Ⓓ	136) Ⓐ Ⓑ Ⓒ Ⓓ
17) Ⓐ Ⓑ Ⓒ Ⓓ	57) Ⓐ Ⓑ Ⓒ Ⓓ	97) Ⓐ Ⓑ Ⓒ Ⓓ	137) Ⓐ Ⓑ Ⓒ Ⓓ
18) Ⓐ Ⓑ Ⓒ Ⓓ	58) Ⓐ Ⓑ Ⓒ Ⓓ	98) Ⓐ Ⓑ Ⓒ Ⓓ	138) Ⓐ Ⓑ Ⓒ Ⓓ
19) Ⓐ Ⓑ Ⓒ Ⓓ	59) Ⓐ Ⓑ Ⓒ Ⓓ	99) Ⓐ Ⓑ Ⓒ Ⓓ	139) Ⓐ Ⓑ Ⓒ Ⓓ
20) Ⓐ Ⓑ Ⓒ Ⓓ	60) Ⓐ Ⓑ Ⓒ Ⓓ	100) Ⓐ Ⓑ Ⓒ Ⓓ	140) Ⓐ Ⓑ Ⓒ Ⓓ
21) Ⓐ Ⓑ Ⓒ Ⓓ	61) Ⓐ Ⓑ Ⓒ Ⓓ	101) Ⓐ Ⓑ Ⓒ Ⓓ	141) Ⓐ Ⓑ Ⓒ Ⓓ
22) Ⓐ Ⓑ Ⓒ Ⓓ	62) Ⓐ Ⓑ Ⓒ Ⓓ	102) Ⓐ Ⓑ Ⓒ Ⓓ	142) Ⓐ Ⓑ Ⓒ Ⓓ
23) Ⓐ Ⓑ Ⓒ Ⓓ	63) Ⓐ Ⓑ Ⓒ Ⓓ	103) Ⓐ Ⓑ Ⓒ Ⓓ	143) Ⓐ Ⓑ Ⓒ Ⓓ
24) Ⓐ Ⓑ Ⓒ Ⓓ	64) Ⓐ Ⓑ Ⓒ Ⓓ	104) Ⓐ Ⓑ Ⓒ Ⓓ	144) Ⓐ Ⓑ Ⓒ Ⓓ
25) Ⓐ Ⓑ Ⓒ Ⓓ	65) Ⓐ Ⓑ Ⓒ Ⓓ	105) Ⓐ Ⓑ Ⓒ Ⓓ	145) Ⓐ Ⓑ Ⓒ Ⓓ
26) Ⓐ Ⓑ Ⓒ Ⓓ	66) Ⓐ Ⓑ Ⓒ Ⓓ	106) Ⓐ Ⓑ Ⓒ Ⓓ	146) Ⓐ Ⓑ Ⓒ Ⓓ
27) Ⓐ Ⓑ Ⓒ Ⓓ	67) Ⓐ Ⓑ Ⓒ Ⓓ	107) Ⓐ Ⓑ Ⓒ Ⓓ	147) Ⓐ Ⓑ Ⓒ Ⓓ
28) Ⓐ Ⓑ Ⓒ Ⓓ	68) Ⓐ Ⓑ Ⓒ Ⓓ	108) Ⓐ Ⓑ Ⓒ Ⓓ	148) Ⓐ Ⓑ Ⓒ Ⓓ
29) Ⓐ Ⓑ Ⓒ Ⓓ	69) Ⓐ Ⓑ Ⓒ Ⓓ	109) Ⓐ Ⓑ Ⓒ Ⓓ	149) Ⓐ Ⓑ Ⓒ Ⓓ
30) Ⓐ Ⓑ Ⓒ Ⓓ	70) Ⓐ Ⓑ Ⓒ Ⓓ	110) Ⓐ Ⓑ Ⓒ Ⓓ	150) Ⓐ Ⓑ Ⓒ Ⓓ
31) Ⓐ Ⓑ Ⓒ Ⓓ	71) Ⓐ Ⓑ Ⓒ Ⓓ	111) Ⓐ Ⓑ Ⓒ Ⓓ	
32) Ⓐ Ⓑ Ⓒ Ⓓ	72) Ⓐ Ⓑ Ⓒ Ⓓ	112) Ⓐ Ⓑ Ⓒ Ⓓ	
33) Ⓐ Ⓑ Ⓒ Ⓓ	73) Ⓐ Ⓑ Ⓒ Ⓓ	113) Ⓐ Ⓑ Ⓒ Ⓓ	
34) Ⓐ Ⓑ Ⓒ Ⓓ	74) Ⓐ Ⓑ Ⓒ Ⓓ	114) Ⓐ Ⓑ Ⓒ Ⓓ	
35) Ⓐ Ⓑ Ⓒ Ⓓ	75) Ⓐ Ⓑ Ⓒ Ⓓ	115) Ⓐ Ⓑ Ⓒ Ⓓ	
36) Ⓐ Ⓑ Ⓒ Ⓓ	76) Ⓐ Ⓑ Ⓒ Ⓓ	116) Ⓐ Ⓑ Ⓒ Ⓓ	
37) Ⓐ Ⓑ Ⓒ Ⓓ	77) Ⓐ Ⓑ Ⓒ Ⓓ	117) Ⓐ Ⓑ Ⓒ Ⓓ	
38) Ⓐ Ⓑ Ⓒ Ⓓ	78) Ⓐ Ⓑ Ⓒ Ⓓ	118) Ⓐ Ⓑ Ⓒ Ⓓ	
39) Ⓐ Ⓑ Ⓒ Ⓓ	79) Ⓐ Ⓑ Ⓒ Ⓓ	119) Ⓐ Ⓑ Ⓒ Ⓓ	
40) Ⓐ Ⓑ Ⓒ Ⓓ	80) Ⓐ Ⓑ Ⓒ Ⓓ	120) Ⓐ Ⓑ Ⓒ Ⓓ	